FAA-AEE-04-01
DTS-34-FA34T-LR3
Federal Aviation Administration
Office of Environment and Energy
Washington, D.C. 20591

FINAL REPORT:

THE USE OF LIDAR TO CHARACTERIZE AIRCRAFT

INITIAL PLUME CHARACTERISTICS

Roger L. Wayson
Gregg G. Fleming
Brian Kim
U.S. Department of Transportation
Research and Special Programs Administration
John A. Volpe National Transportation Systems Center
Environmental Measurement and Modeling Division, DTS-34
Air Quality Facility, Kendall Square
Cambridge, MA 02142-1093

Wynn L. Eberhard
W. Alan Brewer
NOAA Environmental Technology Laboratory
Optical Remote Sensing Division

Letter Report

February 2004

U.S. Department of Transportation
Federal Aviation Administration

Final Report: The Use of LIDAR to Characterize Aircraft Initial Plume Characteristics

EXECUTIVE SUMMARY

A Federal Aviation Administration's (FAA) Emission and Dispersion Modeling System (EDMS) study was conducted from May 17 to 24, 2001, with the goal of measuring the initial plume characteristics of jet exhaust plumes. These initial plume characteristics include plume rise, horizontal plume standard deviation, and vertical plume standard deviation. This data is needed as input to dispersion models for use in complying with air quality requirements at airports. Very little research had been done in this area, and input values previously used in the EDMS were primarily based on engineering judgment. With the improvements being made to EDMS, it became apparent that greater detail was needed for these data to continue the improvement in estimating local pollutant concentrations.

The opportunity to conduct such a study presented itself when a major airport, Los Angeles International Airport (LAX), needed similar data on plume characteristics. The FAA, with support from the Volpe Center, took the lead in the joint effort to identify the best technique for gathering the data, conducting the study, and evaluating the results.

LIDAR (LIght Detection And Ranging) equipment was chosen as the measurement technique. A LIDAR system transmits a pulse of light in a narrow beam and detects the backscatter from light scattering particles as a function of time. The concentration of light scattering particles as a function of range along the pulse's path can be inferred from the time series of detected light intensity. By scanning with the LIDAR in a defined direction over a period of time with many LIDAR pulses, the distribution of particles over the region of the sweep (e.g., a vertical plane or plume cross section) can be determined. LIDAR was considered well suited for measuring the geometry of plumes which contain light-scattering particles as the tracer. Cross-sections of the plume were measured at a variety of distances behind the aircraft during takeoff roll. This final study report is based on an analysis of 4138 LIDAR sweeps, or cross sections, collected at LAX.

The National Oceanic and Atmospheric Administration (NOAA) supplied and operated two LIDAR units for this research. One system used ultraviolet light as the scan laser and was called the *Ozone Profiling Atmospheric LIDAR (OPAL)*. The second LIDAR unit used the infrared spectrum and was named the *High Resolution Solid State Doppler LIDAR (HRDL)*. Each unit was contained in a trailer and set in place near the active runway and taxiway.

The OPAL system, operating at the ultraviolet wavelength of 0.355 μm (after slight modification for this study to measure only aerosol backscatter and not generate the additional wavelengths used to measure ozone), proved to be the more effective of the two systems in determining the plume parameters. This was expected since the aerosols/particulate matter emitted by the airplane is what is causing the back scatter of energy from the LIDAR laser and are "seen" as the plume. These components are very small, typically less than 100 nanometers with an average diameter of about 30 nanometers. The HRDL system, using infra-red frequencies with a wavelength of 2.02 μm was not as sensitive to the very small aerosol size.

Concurrent to the sampling by the LIDAR units, a spotter was used to identify aircraft as the measurements occurred. This included the aircraft type, airline, and the tail number during the daylight hours. At night it was not always possible to read the tail numbers and only the aircraft type and airline were consistently reported. Multiple still pictures and filming of events were also performed. Requests were also made for tower operational data and weather data collected at the airport. These data were integrated into the final database.

It can be concluded from the measurements that significant plume rise occurs for the jet/turbine exhaust plume. It can also be concluded that initial plume spread is significant and greater than previously thought. Findings in this report represent aggregate values for plume rise and initial plume standard deviations. One set of data is recommended for both large commercial aircraft and smaller commuter aircraft until more data become available. These final parameters suggested for use are:

$$\text{Sigma Y} = 10.5 \text{ meters}$$
$$\text{Sigma Z} = 4.1 \text{ meters}$$
$$\text{Plume Rise} = 12 \text{ meters}$$

Effects of temperature, wind speed, wind direction, and turbulence (stability class) were not found to be statistically significant in the data analysis. While differences do appear to occur by aircraft types, it cannot be proven they are significantly different at this time and more measurements are needed to follow up on this trend analysis.

The results of the measurements would tend to support that jet exhaust plume rise occurs due to the plume's initial thermal buoyancy. Plume rise has been studied intensely for stationary sources such as stacks, and semi-empirical equations have been developed and successfully applied in predicting plume rise for stationary sources. The important variables used in these approaches often include the wind speed, downwind distance, and heat emission rate. However, jet exhaust plumes have not been studied to the same extent. In analyzing the jet exhaust plume data collected in this study, no correlation was found for wind speed or downwind distance. This implies that thermal buoyancy is the overriding variable for the plume rise of a jet/turbine engine exhaust. This is an important finding that should be further evaluated when more data becomes available.

In summary, it can be concluded that significant plume rise occurs for the jet/turbine exhaust, and that initial plume spread is significant and greater than previously thought. Since there is only one data set, study results for plume rise and initial plume standard deviations were calculated using a conservative basis (i.e., the plume was only measured to well-defined boundaries, the plume rise was based on the sweep corresponding to the second greatest height or elevation measured, the sweep used for the plume rise was also used to determine the standard deviation of the plume). Effects of aircraft type, temperature, wind speed, wind direction, or turbulence (stability class) were not found to be statistically significant in the data analysis. These potential factors must be further explored when more data become available.

Final Report: The Use of LIDAR to Characterize Aircraft Initial Plume Characteristics

This study provides new insights into aircraft plume behavior that greatly surpasses historic understanding, and data for more accurate modeling of plume rise and spread from commercial aircraft at airports. This final report completes individual analysis of the LAX data set, initially reported in the related *Preliminary Report* published in September 2002. Additional studies are planned (based on available funding) to analyze potential changes in the derived parameters due to site characteristics (e.g., elevation, weather conditions) and will be reported on as the work continues.

Final Report: The Use of LIDAR to Characterize Aircraft Initial Plume Characteristics

TABLE OF CONTENTS

EXECUTIVE SUMMARY 2

INTRODUCTION 6

APPROACH 6

MEASUREMENT BACKGROUND 8

DATA REDUCTION AND ANALYSIS 10

DERIVATION OF INITIAL PLUME PARAMETERS 12

STATISTICAL TESTING 13

    Aircraft Type Analysis 13

    Meteorological Analysis 17

        Temperature Effects..18
        Wind Speed Effects..18
        Wind Direction Effects..18
        Turbulence 18

DATA RESULTS 29

CONCLUSIONS 29

    General 29

    Plume Rise Dependence on Thermal Buoyancy 30

FUTURE WORK 34

INTRODUCTION

A Federal Aviation Administration's (FAA) Emission and Dispersion Modeling System (EDMS) study was initiated in the Spring of 2001 with the goal of measuring the initial plume characteristics of jet exhaust plumes. These initial plume characteristics include plume rise, horizontal plume standard deviation, and vertical plume standard deviation. This data is needed as input to dispersion models for use in complying with air quality requirements at airports. Very little research had been done in this area, and input values previously used in the EDMS were primarily based on engineering judgment. With the improvements being made to EDMS, it became apparent that greater detail was needed for these data to continue the improvement in estimating local pollutant concentrations.

The opportunity to conduct such a study presented itself when a major airport, Los Angeles International Airport (LAX), needed similar data on plume characteristics. The FAA and LAX decided to conduct a joint study on initial plume characteristics. The FAA, with support from the Volpe Center, took the lead in the joint effort to identify the best technique for gathering the data, conducting the study, and evaluating the results.

This initial aircraft plume behavior study was conducted at LAX from May 17 to 24, 2001, using LIDAR (LIght Detection And Ranging) equipment. Cross-sections of the plume were measured at a variety of distances behind the aircraft during takeoff roll. The initial behavior was expected to depend on aircraft characteristics, including physical size of the engines and their position on the airframe. Initial plume behavior was also expected to vary somewhat with atmospheric stability and local wind conditions. Dispersion is inherently a random process, so many experimental cases are required to determine the mean behavior and typical variability. This final study report is based on an analysis of 4138 LIDAR sweeps, or cross sections, collected at LAX. This study provides new insights into aircraft plume behavior, and data for more accurate modeling of plume rise and spread from commercial aircraft at airports. This final report completes individual analysis of the LAX data set, initially reported in the related *Preliminary Report* published in September 2002. Additional studies are planned (based on available funding) to analyze potential changes in the derived parameters due to site characteristics (e.g., elevation, weather conditions) and will be reported on as the work continues.

APPROACH

It was originally envisioned that the initial dispersion parameters could be measured by using an instrumented tower near an active runway and/or taxiway. It soon became apparent that this would not be feasible because of safety concerns. Alternative measurement schemes were evaluated, with the most promising being the use of LIDAR (LIght Detection And Ranging).

A LIDAR system transmits a pulse of light in a narrow beam and detects the backscatter from light scattering particles as a function of time. The concentration of light scattering particles as a function of range along the pulse's path can be inferred from the time series of detected light intensity. By scanning with the LIDAR in a defined direction over a period of time with many LIDAR pulses, the distribution of particles over the region of the sweep (e.g., a vertical plane or plume cross section) can be determined. LIDAR was considered well suited for measuring the geometry of plumes containing light-scattering particles as the tracer.

Although the use of LIDAR to characterize aircraft plumes is a new concept, LIDAR has been used for several pollution studies in the past. Examples of these studies include: (1) an EPA study on remote sensing of automobile emissions[1], (2) work by the University of Iowa/Los Alamos National Laboratory to investigate traffic particulate emissions[2], (3) work by the South Coast/Southeast Desert Air Basin for pollution transport using an EPA aircraft mounted LIDAR system[3], and (4) work by the University of Colorado/NOAA for power plant plumes[4].

The study conducted by EPA stated that LIDAR is an excellent way to do remote sensing for the plume from automobile exhaust. Also, the report concluded, "The UV Raman LIDAR system is expected to enable a variety of commercial environmental monitoring products, including automobile emissions monitoring, light and heavy duty truck emissions monitoring, aircraft emission monitoring, warning systems for toxic chemical spills, and fence line monitoring."

In the study performed by the University of Iowa/Los Alamos National Laboratory, the shape of exhaust plumes from motor vehicle traffic was captured using LIDAR. Since the use of LIDAR is relatively new for this type of application, established data analysis and reduction procedures do not exist. However, this study stated that the shape of these plumes can be determined once data analysis and reduction procedures are defined for the individual application.

The study in the South Coast of California pointed out that if methods were developed, LIDAR could be used to make quick effective testing for pollutant transport. Many examples also exist for stationary sources, such as the study by the University of Colorado/NOAA. In the study, LIDAR was used to measure the plume characteristics from stationary sources (as was done in this report for aircraft). Although only four example reports are discussed above, there are many other reports that include similar details and results. These various studies point out that LIDAR is a viable remote sensing tool for plume characterization. Therefore, it was thought that LIDAR could be used to

---

[1] U.S. EPA, Remote Sensing of Automobile Emissions using Raman LIDAR, Contract Number 68D00262 http://cfpub.epa.gov/ncer_abstracts/index.cfm/fuseaction/display.abstractDetail/abstract/1704, Project Dates September 1, 2000 through March 1, 2001.
[2] University of Iowa/ Los Alamos National Laboratory, Measurement of Traffic Particulate Emissions and Incident Detection, www.iihr.uiowa.edu/projects/new_jersey .
[3] California Air Resources Board, Utilization of Remote Sensing Data In the Evaluation of Air Pollution Characteristics in the South Coast/Southeast Desert Air Basin, www.arb.ca.gov/research/abstracts/a2-106-32.htm.
[4] University of Colorado / NOAA, Airborne lidar characterization of power plant plumes during the 1995 Southern Oxidants Study, www.agu.org/pubs/abs/jd/98JD02625/tmp.html, 1995.

measure plume parameters of jet aircraft, and methods could be derived to accomplish this goal from the raw collected data.

In support of AEE, the Volpe Center then initiated a search to identify the most qualified organization to provide LIDAR support. After contacting NASA, major universities, and private industry, it became apparent that the most qualified organization for this study was the National Oceanic and Atmospheric Administration (NOAA). NOAA has several LIDAR units and the flexibility to re-engineer the units and associated software on a project-by-project basis.

MEASUREMENT BACKGROUND

LIDAR was used in this study to observe the time-varying position and geometry of the jet exhaust. Two LIDAR units were used in this research. One system used ultraviolet light as the scan laser and was called the *Ozone Profiling Atmospheric LIDAR (OPAL)*. The second LIDAR unit used the infrared spectrum and was named the *High Resolution Solid State Doppler LIDAR (HRDL)*. Each unit was contained in a trailer and set in place near the active runway and taxiway. Figure 1 shows a picture of the OPAL system trailer.

Figure 1. The OPAL System

The OPAL system, operating at the ultraviolet wavelength of 0.355 μm (after slight modification for this study to measure only aerosol backscatter and not generate the additional wavelengths used to measure ozone), proved to be the more effective of the two systems in determining the plume parameters. This was expected since the aerosols/particulate matter emitted by the airplane are what is causing the back scatter of energy from the LIDAR laser and are "seen" as the plume. These components are very small, typically less than 100 nanometers with an average diameter of about 30 nanometers. The HRDL system, using infra-red frequencies with a wavelength of 2.02 μm was not as sensitive to the very small aerosol size. The OPAL system, using a smaller wavelength, was more easily able to "visualize" the plume. As such, only the OPAL system was used for this study and only the data from this system is discussed in this report. Airplanes on both the taxiway at idle and on a runway during initial take-off roll were measured. Figure 2 shows an example output from the OPAL system.

Figure 2. Example Output from the OPAL System

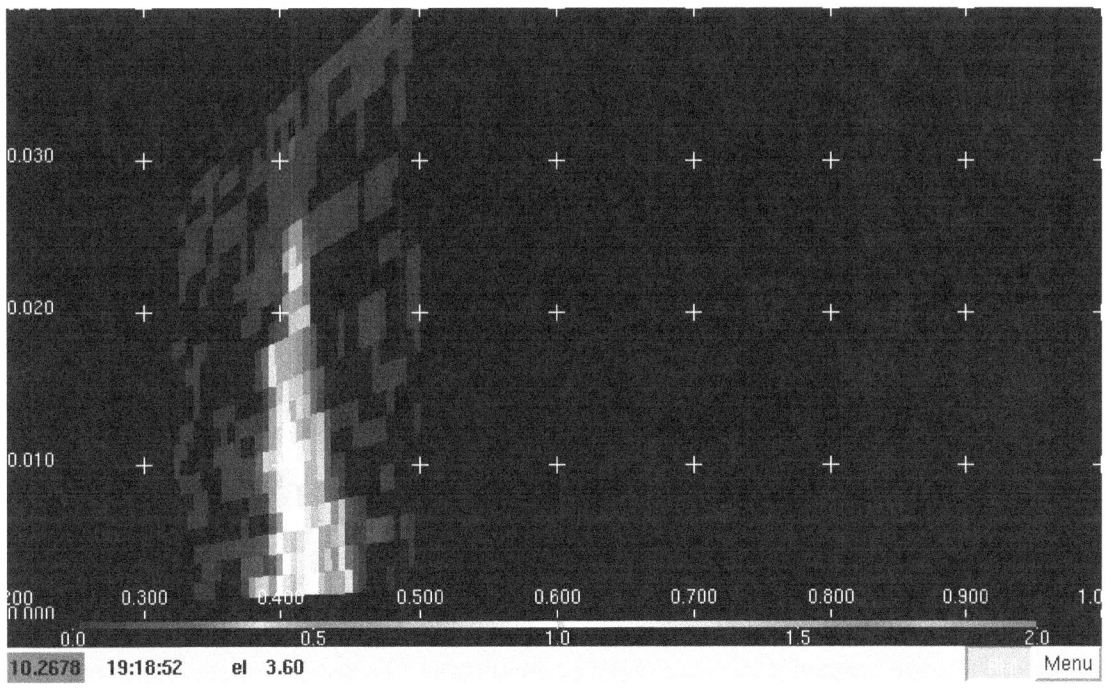

Additionally, only one path (or the cross sections measured at one stationary position behind the aircraft) is presented in this report. This path, or horizontal plane of the LIDAR sweep, is shown in Figure 3 as the red line. If additional funding can be identified for NOAA data reduction, subsequent reports may include data (such as wind movements) from the infrared scanning LIDAR (HRDL), which collected data at the other sweep angles shown in Figure 3.

Figure 3: Sweep Angles For LIDAR Units
(Only the Red Sweep is Discussed in This Report)

Concurrent to the sampling by the LIDAR units, a spotter was used to identify aircraft as the measurements occurred. This included the aircraft type, airline, and the tail number during the daylight hours. At night it was not always possible to read the tail numbers and only the aircraft type and airline were consistently reported. Multiple still pictures and filming of events were also performed. Requests were also made for tower operational data and weather data collected at the airport. These data were integrated into the final database.

DATA REDUCTION AND ANALYSIS

After data collection, initial quality control was performed by NOAA personnel to eliminate any anomalies. Computer graphics were then created to illustrate the results of the sweeps in characterizing the plume. Figure 4 shows such an example illustration of a sweep. In the Figure 4 example, the center of the plume can be easily identified by the area of greatest concentration (red center). This represents the height of the plume, or plume rise. The outer boundaries of the plume are also easily identified allowing the width and height of the plume to be quantified.

For each sweep (cross section) care was then taken by Volpe personnel to carefully determine the height of the center of the plume (Z_{center}), which represents the plume rise. The outer boundaries of the plume were also carefully determined for each sweep. The scale on the side of the chart should also be noted. To make sure that the plume was represented accurately, and not other concentrations or interferences, the plume was only measured to the well-defined boundaries. In the Figure 4 example, this was to the light brown (i.e., 0.54 ρ_{pl} / ρ_{bkg}) fringe as shown. Since the scale is the measured density or concentration of aerosol to the background (ρ_{pl} / ρ_{bkg}) the ratio of concentrations from the center (red) to defined outer fringe could be determined since:

$$(\rho_{plc} / \rho_{bkg}) / (\rho_{plf} / \rho_{bkg}) = (\rho_{plc} / \rho_{plf}) \qquad [1]$$

where: ρ_{plc} = density (concentration) of plume center
ρ_{plf} = density (concentration) of plume fringe

Figure 4. Example of Computer Enhanced LIDAR Image

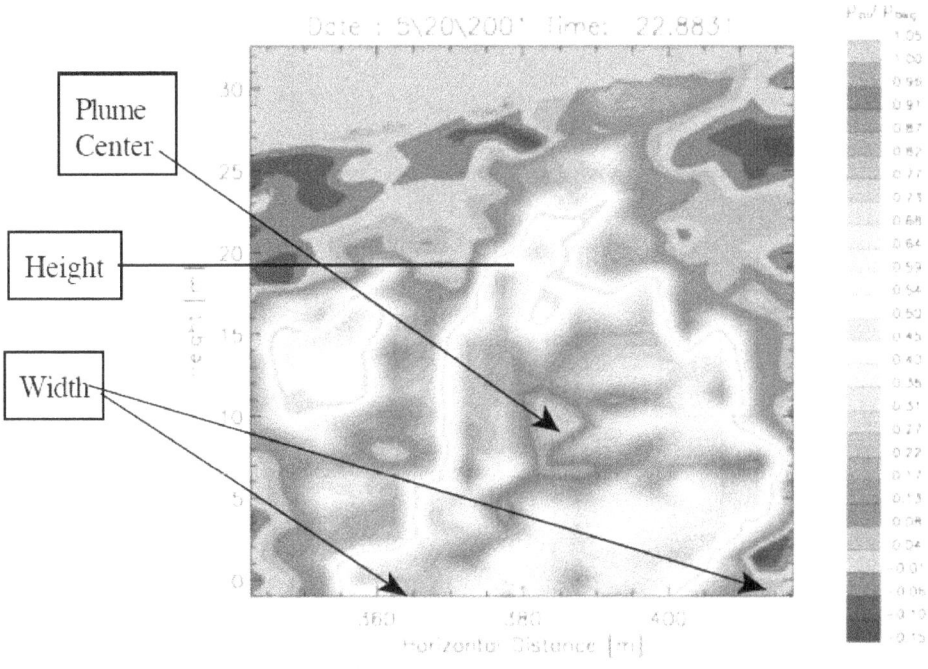

Unfortunately, not all sweeps were as easy to determine the center and fringes as the example shown. The example is a well-behaved plume. Plume break-up, multiple centers, high plume rise, and irregular shapes required careful judgment in many cases. The only way to accurately determine the plume was to review the time series of events (sweeps) behind the aircraft and envision the plume in total, which was done for all sweeps by Volpe. This process was difficult due to the effect of the high velocity jet from the aircraft, which tended to cause irregular mixing and false centers of the actual plume. These data were carefully reduced in such a way to allow for a more systematic recreation of a symmetric plume in the final analysis. This recreation provides added quality assurance of the data presented in this report.

Once the center of the plume and the plume fringe were determined for each sweep used in the analysis, the plume rise for the aircraft event was determined. The plume rise was based on the sweep corresponding to the second greatest height or elevation measured and used in the final plume rise determination. This approach provided a level of conservatism as compared with selecting the sweep with the greatest height or plume rise. A conservative approach was also used to determine the standard deviation of the plume. For the standard deviation determination, the same sweep used for the plume rise was used. The ratio of the concentrations was determined as well as the distance to the fringe from the center of the plume. Using this information, and the basic Gaussian equation, the instantaneous standard deviation was determined. It should be noted that the Gaussian models use time-averaged standard deviations, but a starting point of the plume must be determined and used during modeling. These instantaneous sweeps provide the information to allow this starting point to be determined. All data was then included in a commercially available spreadsheet and the initial plume parameters derived.

Statistical testing of the data was also used to determine if variance in the weather or the type of aircraft resulted in correlated changes with the determined plume parameters. This testing was done by first combining key measured data with reported meteorological data reported by NOAA. Once derived, this large spreadsheet database permitted sorting, analysis, and statistical testing of various data set combinations. Results of this work are included in this report.

DERIVATION OF INITIAL PLUME PARAMETERS

The derived initial plume parameters are based on all of the useable data collected. From this data (4138 sweeps) events were characterized for each aircraft event. As previously stated, the second highest value for plume rise was then selected to allow a conservative estimate. This final data base included 380 events for large commercial aircraft and 49 events for commuter aircraft.

The results of this analysis were first done in the aggregate for all large commercial aircraft events and separately for commuter aircraft as was documented in the September 2002 *Preliminary Report*. Derived parameters included the plume rise, the standard deviation of the plume spread in the horizontal (Sigma Y) and the standard deviation of the plume spread in the vertical (Sigma Z). The results of the analysis are:

Large Commercial Aircraft (primarily turbofan engines)

    Sigma Y =   10.8 meters
    Sigma Z =   4.1 meters
    Plume Rise =   11.9 meters

Commuter Aircraft (primarily turboprops)

    Sigma Y =   10.3 meters
    Sigma Z =   4.1 meters
    Plume Rise =   12.1 meters

Of great interest and somewhat of a surprise to the research team were the small differences between the two categories of aircraft. From this aggregate analysis, testing shows us that there are no statistical differences in the two distribution populations. As such, the same value of plume rise and plume standard deviations could be used for both types of aircraft. This has been done and is included in the conclusion section of this report.

STATISTICAL TESTING

Additional questions remained about the collected data. These included the effect on plume rise and plume spread due to the variances of large commercial aircraft engine location and/or the effects of weather. The first step in this analysis was the creation of a master spreadsheet that included key pieces of information from the very large raw data spreadsheets. Statistical testing was then used for two purposes: 1) to further sub-divide the data into aircraft types to see if aircraft size and/or engine mounting location caused any significant variance in the derived parameters; and, 2) to determine the significance of the effects of weather (wind speed, wind direction, stability class and temperature) on the derived parameters. This section presents these results.

Aircraft Type Analysis

To begin the aircraft type analysis, a scattergram was created to allow visual comparisons of the plume spread and plume rise for each aircraft event. Figure 5 shows the graphical comparison. The x-axis (abscissa) is the value of the total plume spread while the y-axis (ordinate) is the center of the plume after plume rise has occurred. It should be noted that plume spread, and not standard deviation, is shown in Figure 5. This is because this was the parameter directly measured (see Figure 4). Since the standard deviation is derived from this measured parameter, statistical results will follow the same trends, albeit with a different abscissa scale. Figure 6 shows the same comparison as Figure 5, but with standard deviations used instead of plume spread. It can be seen the plots follow the same trends. As such, to avoid errors in calculations and to follow good practices of working with directly measured data, plume spread was evaluated in further testing.

Large commercial aircraft and the smaller commuter aircraft are shown as different colors and symbols in Figure 5 and 6. It can be seen that the commuter aircraft is inter-dispersed throughout the same region occupied by the large commercial aircraft. This visually confirms the aggregate analysis, that the average values and deviation of the large commercial aircraft and smaller commuter aircraft results are very similar.

However, upon closer review of the large commercial aircraft as plotted in Figures 5 and 6, it appears that some events for the lateral dispersion are greater certain aircraft, such as the Boeing 747. The researchers thought this trend could be related to the orientation of the aircraft engine causing the engines to be at different heights, at different distances from the centerline of the fuselage, and near or away from a wing. Other variables could be number of engines, exhaust temperatures, and total exhaust flow. To explore this trend, the data were further explored by breaking the data into specific aircraft types and analyzing the results. Figure 7 shows a comparison of the large commercial aircraft types measured.

Figure 5. Comparison of Large Commercial and Smaller Commuter Aircraft Trends :
Total Plume Spread Vs. Plume Rise

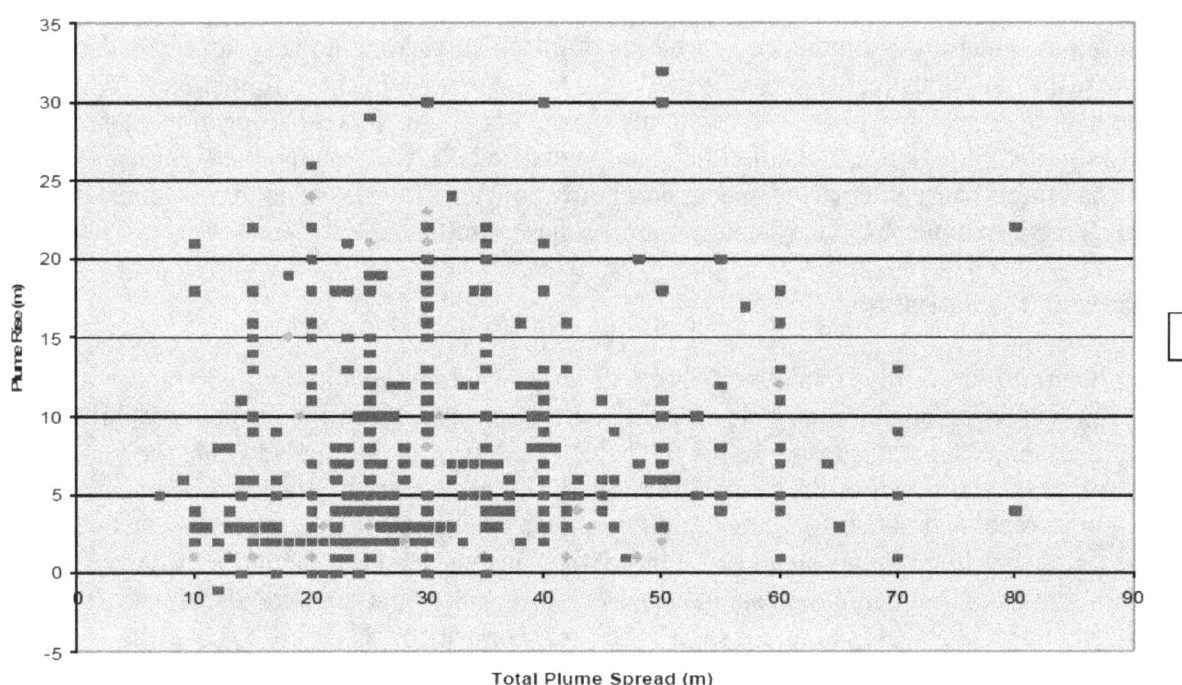

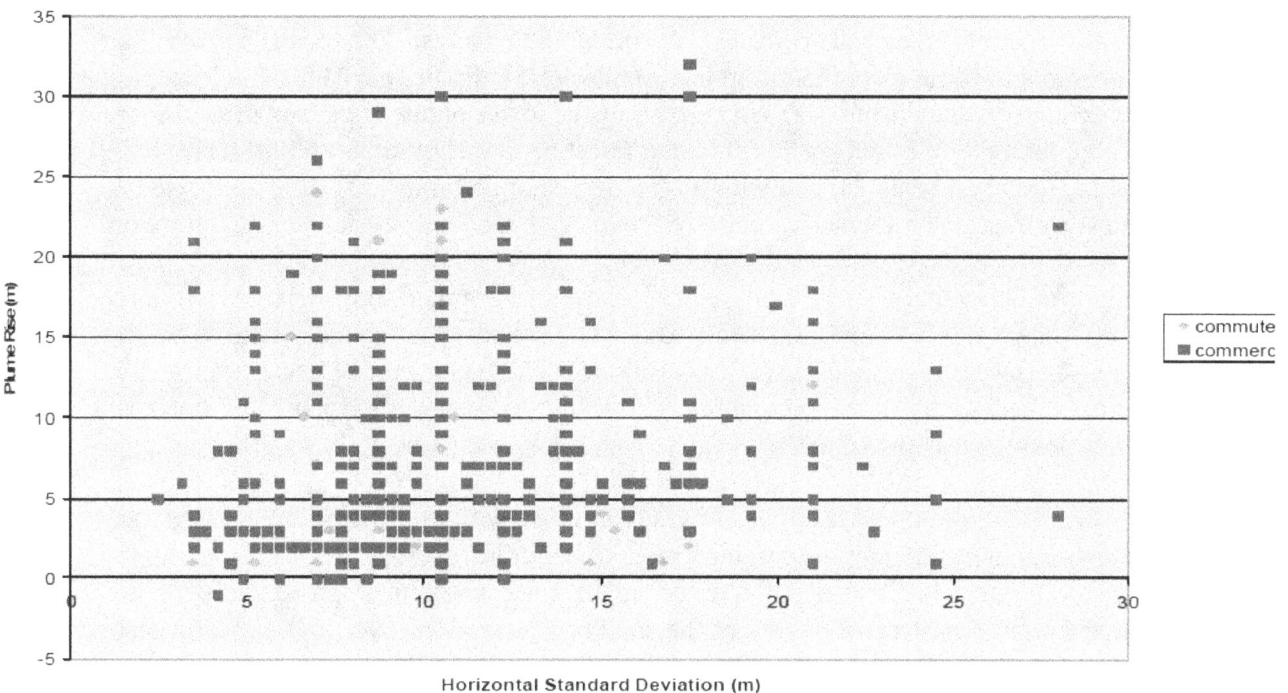

Figure 6. Comparison of Large Commercial and Smaller Commuter Aircraft Trends : Horizontal Standard Deviation Vs. Plume Rise

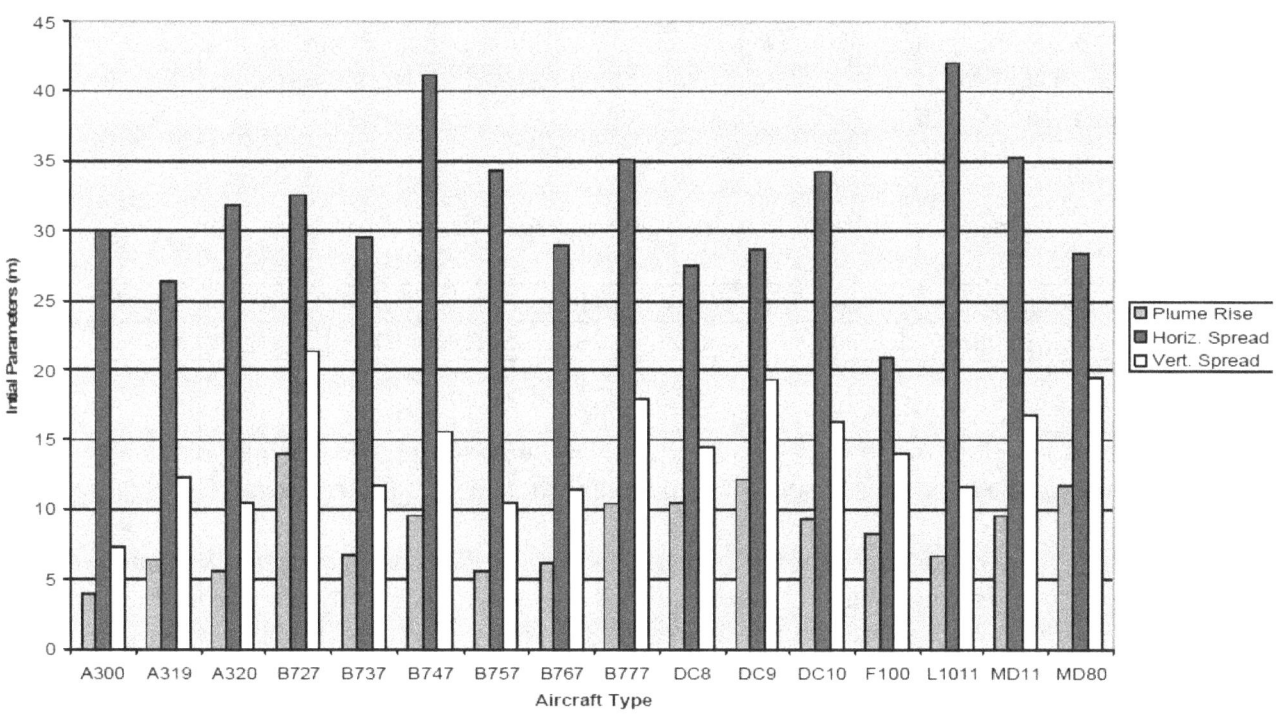

Figure 7. Comparison of Large Commercial Aircraft Initial Plume Parameters

This figure is based on the averages of all aircraft measured in each type. It can be seen that no clear trend exists for all aircraft types but some interesting results do occur. The largest plume rise would appear to be from the Boeing 727. The next two highest plume rises also occur from aircraft with fuselage-mounted engines. The only other fuselage mounted engine aircraft in the group is a smaller plane the Fokker 100. The least plume rise occurs from the Airbus 300. In fact, all of the lower plume rises came from wing mounted aircraft. While these results are interesting and should be used in future research, it cannot be concluded with a high degree of certainty that the plume rise is greatest for fuselage-mounted aircraft. More data is needed to further this assumption.

The largest horizontal width, or plume spread, was determined to occur from two larger aircraft, the Boeing 747 and Lockheed 1011. The smallest horizontal plume spread is from a smaller aircraft, the Fokker 100. However, the other aircraft horizontal plume spread seemed not to be related to aircraft size. For example, the Boeing 767 horizontal plume spread is less than the 757.

The vertical plume spread is similar to the plume rise with the Boeing 727 having the greatest average value and the fuselage mounted aircraft representing the top three values. Similarly, the Airbus 300 had the smallest vertical plume spread and wing mounted aircraft represented some of the smaller values. However, more data is needed to substantiate this finding and other variables need to be further measured and evaluated.

Substantiation of any trend will require more data but values derived from this work at a single airport are summarized in Table 1. The comparison of the overall average parameters of the wing mounted versus fuselage mounted aircraft show numeric differences, but again cannot be substantiated from this single test site. More data is needed to confirm these trends and to understand the reasons for these trends, if they do indeed exist.

Table 1. Comparison of Wing Mounted and Fuselage Mounted Aircraft

Wing Mounted	Plume Rise = 7.6	Horiz. Spread = 33	Vert. Spread = 13
Fuselage Mounted	Plume Rise = 11.6	Horiz. Spread = 28	Vert. Spread = 19

It should be noted that some of the aircraft in the categories included engines in the rear stabilizer. These types were sorted based on the other engines on the aircraft[5].

The analysis for commuter aircraft shows even more variance as displayed in Figure 8. This database is even smaller (e.g., the GA aircraft category had only one event) than the large commercial database and for that reason the aggregate values are thought to be the best approximation at this time until more data is available.

---

[5] The L1011, MD11 and DC10 were considered to be wing mounted although each also includes an engine in the vertical stabilizer in addition to the wing mounted engines.

Meteorological Analysis

Plume rise and plume spread can be influenced by local meteorological variables such as temperature, wind speed, wind direction, and turbulence (stability class). The temperature difference between the ambient air and the jet exhaust provide thermal buoyancy leads to both plume rise and vertical dispersion. The wind speed acts against this vertical motion and may cause the plume rise and vertical plume spread to be reduced while increasing the horizontal spread. The wind direction could have an effect on the initial parameters if it is blowing across the runway as compared to along the runway. Temperature, wind speed, and wind direction have been analyzed directly by comparing the NOAA reported meteorological parameters with the initial plume parameters and are discussed.

The fourth variable, atmospheric stability or turbulence, could not be directly analyzed because the NOAA reported data does not report atmospheric stability. The nighttime measurements, which were taken during very low winds, represent a very stable atmosphere. These events were compared to the daytime sampling events when the atmosphere tended to be more unstable. No real differences were observed and since there are only 15 large commercial aircraft events in this nighttime database, strong confidence could not be placed in this analysis. To develop a more robust database for analysis, the available data was used to determine the Pasquill-Gifford stability classes. This allowed a review of stability using the Pasquill-Gifford classes as the evaluated parameter. This analysis is also included in this report.

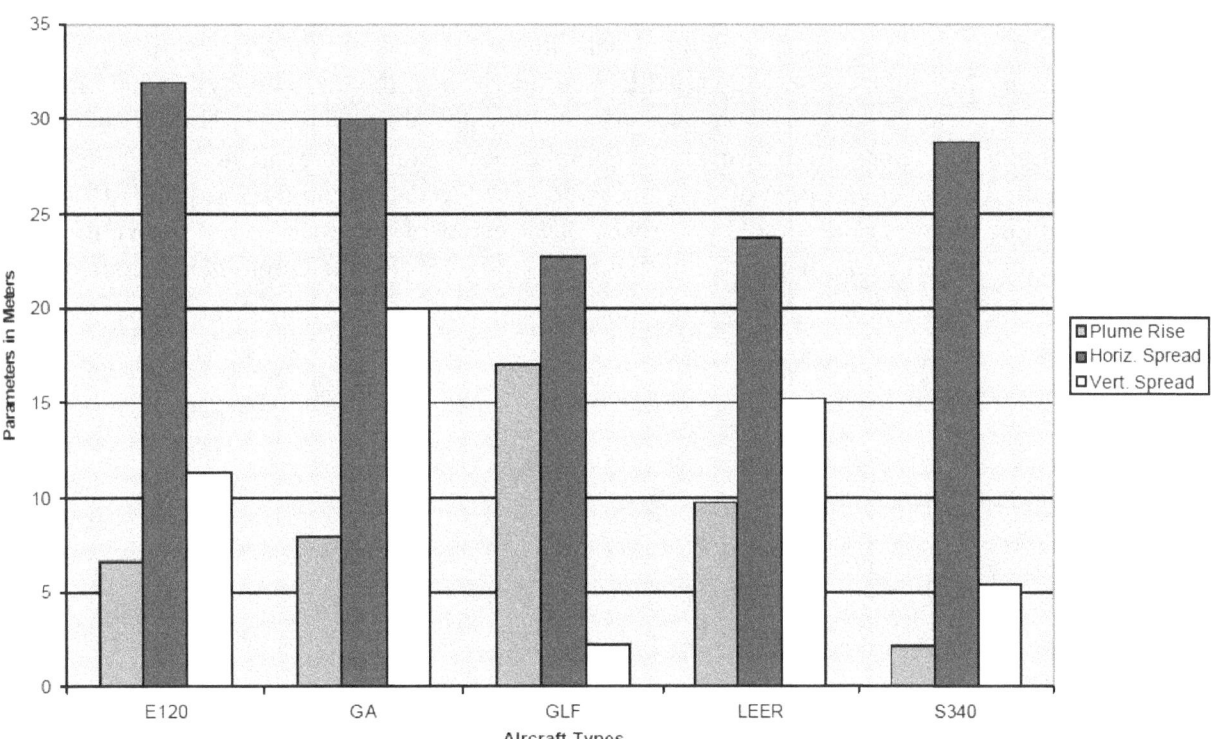

Figure 8. Comparison of Commuter Aircraft Initial Plume Parameters.

Temperature Effects. When temperature was plotted against the three derived initial meteorological parameters (Figures 9 and 10), no significant correlation exists for either the large commercial or the commuter aircraft, as can be seen from the figures. This was for a limited temperature range of temperature, from 16 to 22 degrees Centigrade. The fitted linear trend lines and the resultant R^2 values are presented in each figure (*an R^2 value of 0 represents no correlation while a value of 1 is perfect correlation*). It should be noted that all plots in this section show linear trend lines. During analysis, linear, polynomial, exponential, logarithmic and power curve fits were analyzed. Some provided slightly better fits than did the linear slope, but none were significant. As such, for reader understanding, only the more simple linear correlation trend lines are shown

Wind Speed Effects. Figures 11 and 12 include the results of the wind speed analysis as compared to the three derived initial dispersion parameters. Wind speeds ranged from 0 to 14 knots. Again, no significant trends were apparent.

Wind Direction Effects. The results of the analysis for wind direction are shown in Figures 13 and 14. Wind Direction was essentially out of two primary directions: along the runway (250 degrees) and across the runway diagonally (110 degrees). The runway, 25R, is oriented toward 250 degrees for the end where takeoffs began. Once again, no significant trends were apparent and the trends were even less significant than were the wind speed effects.

Turbulence. A very small amount of data, know to be taken in a very stable atmosphere at night from direct observation, led to a comparison of the three derived initial plume parameters in the known stable and unstable cases. Values were very similar and as such, no trend was apparent. However, for completeness, data were evaluated using the NOAA reported cloud cover, time of day, and wind speed to determine the Pasquill-Gifford Stability Classes as reported by Turner.[6] The results of this analysis are shown in Figure 15a for the large commercial aircraft and in Figure 15b for the commuter aircraft. It can be seen that no trend is apparent.

---

[6] Turner, D.B., Workbook of Atmospheric Dispersion Estimates, U.S. Environmental Protection Agency, Washington, D.C., 1970.

Figure 9. Large Commercial Aircraft Initial Plume Parameters Vs. Ambient Temperature

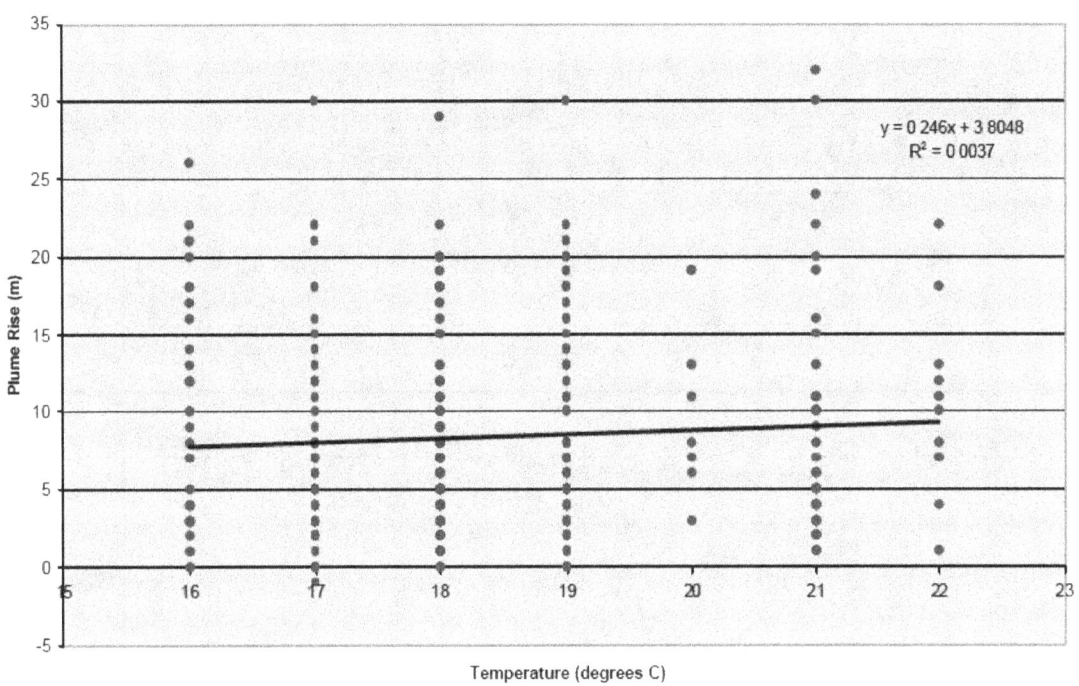

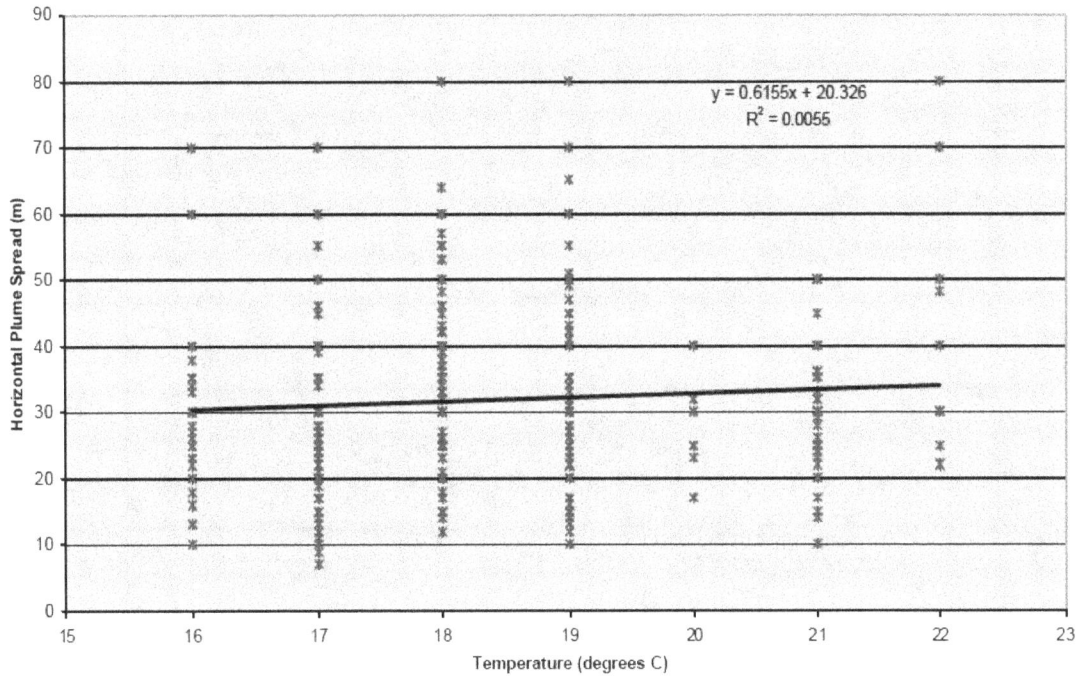

Figure 9. Large Commercial Aircraft Initial Plume Parameters Vs. Ambient Temperature (continued)

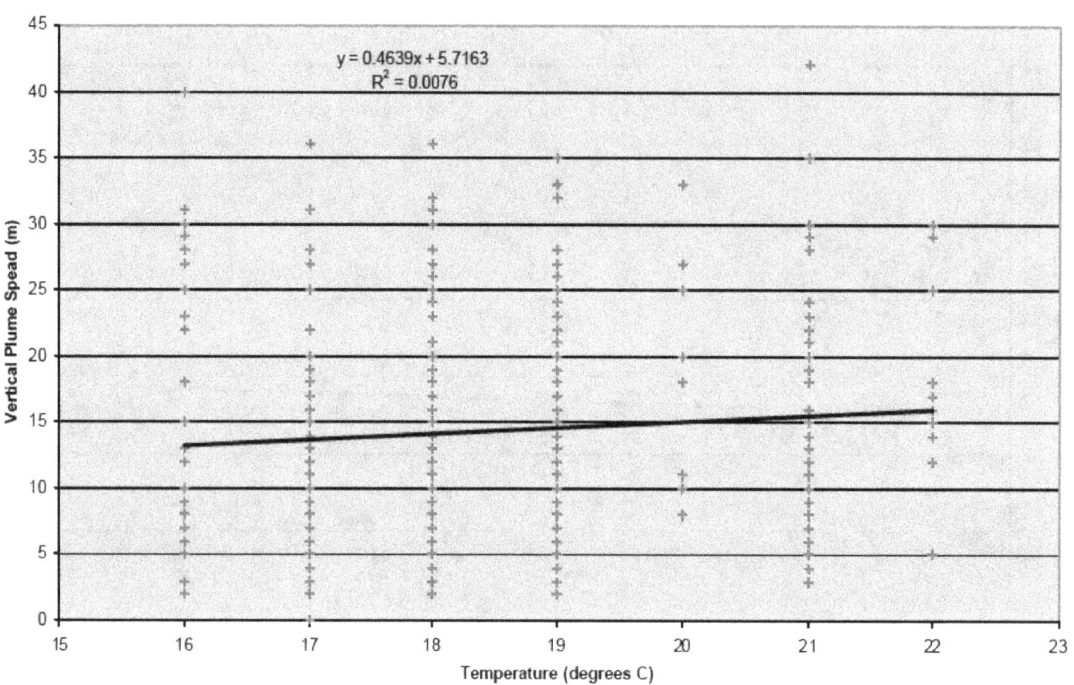

Figure 10. Commuter Aircraft Initial Plume Parameters Vs. Ambient Temperature

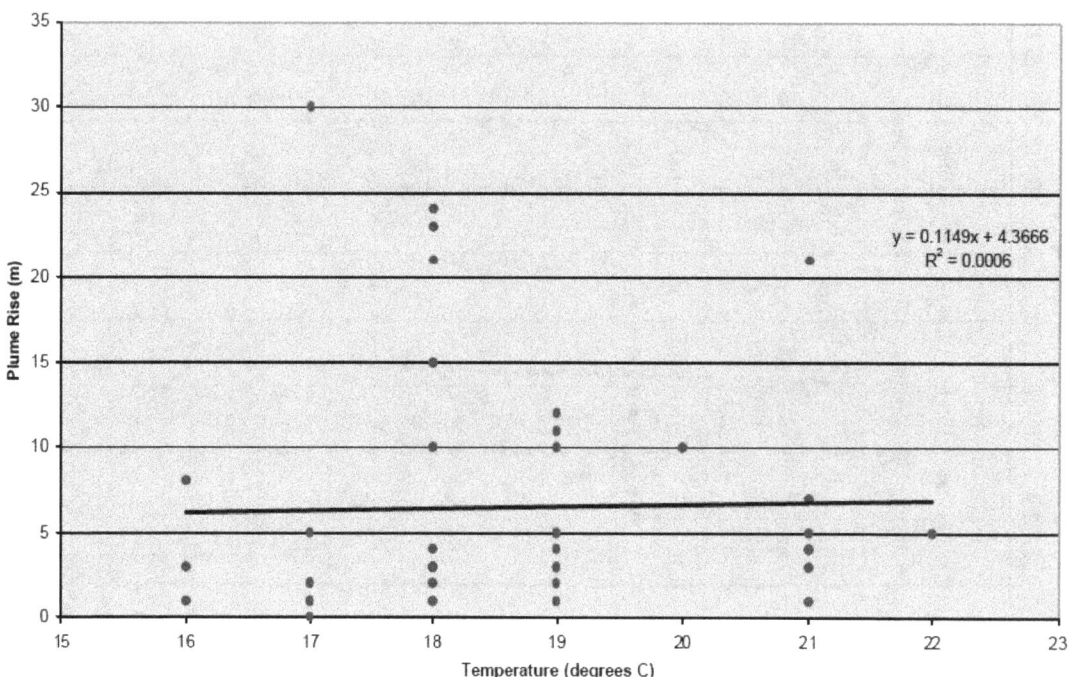

Figure 10. Commuter Aircraft Initial Plume Parameters Vs. Ambient Temperature (continued)

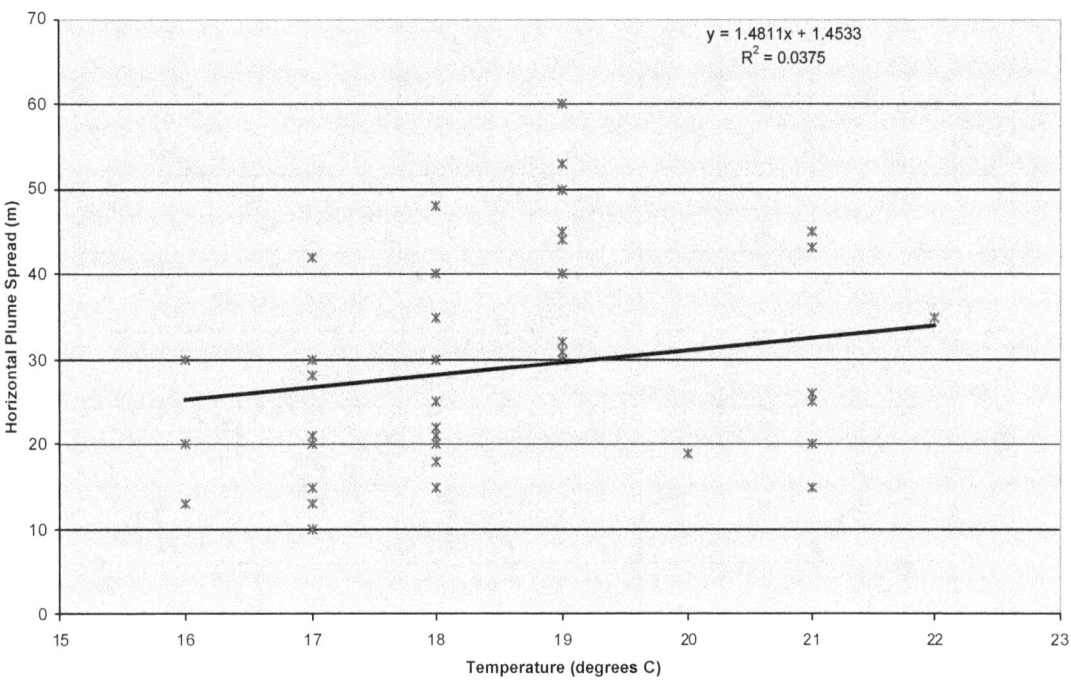

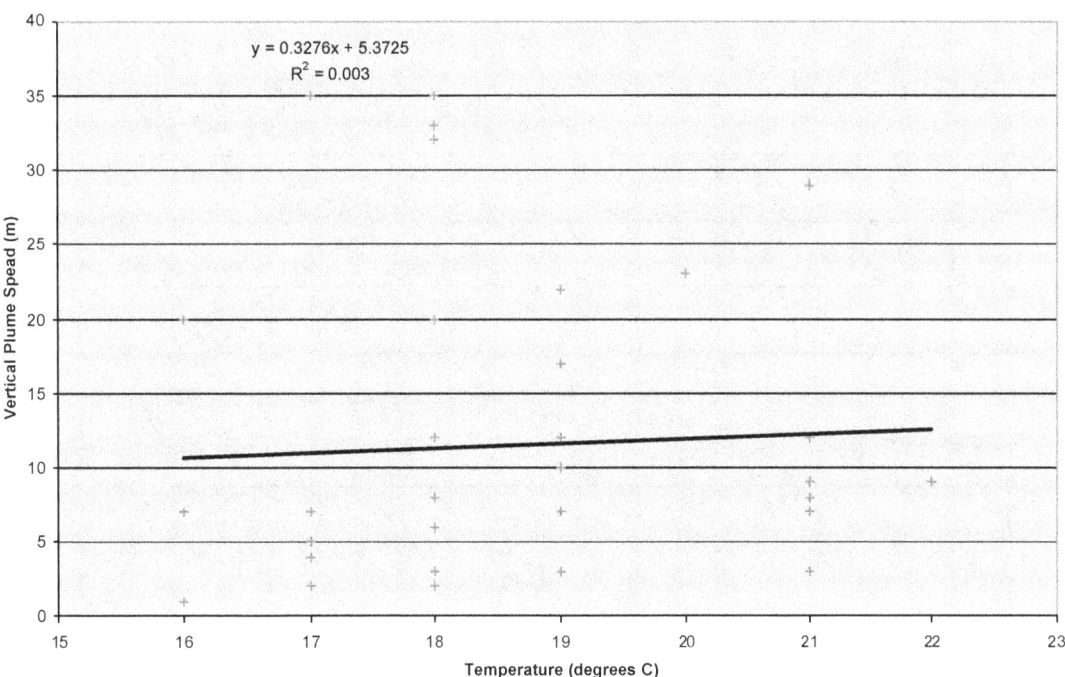

Figure 11. Large Commercial Aircraft Initial Plume Parameters Vs. Wind Speed

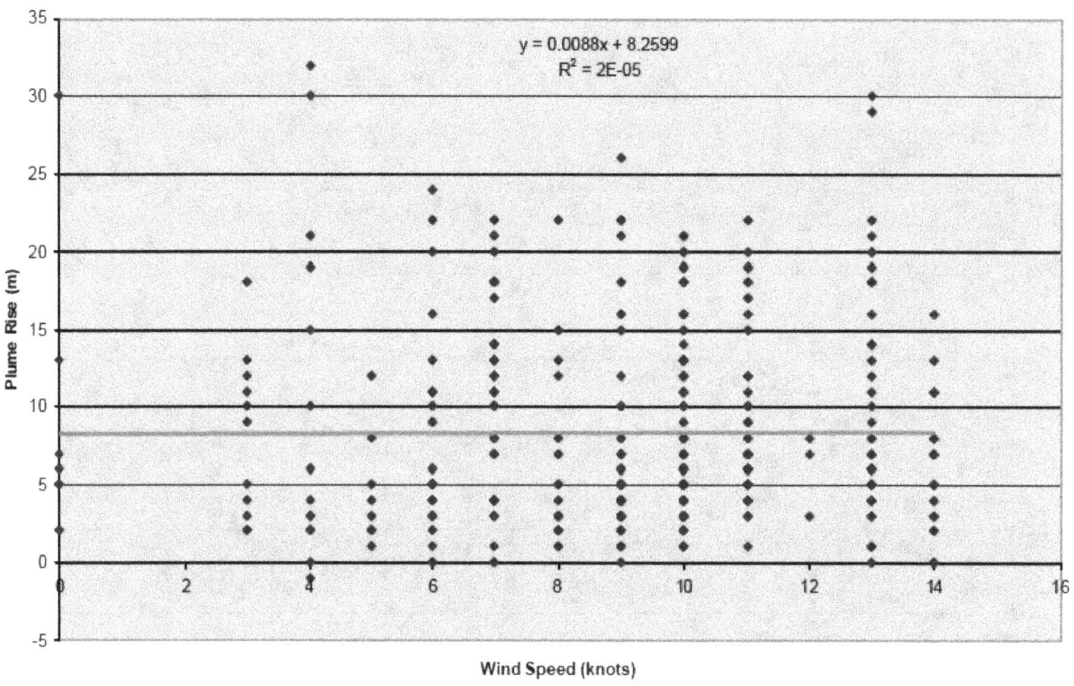

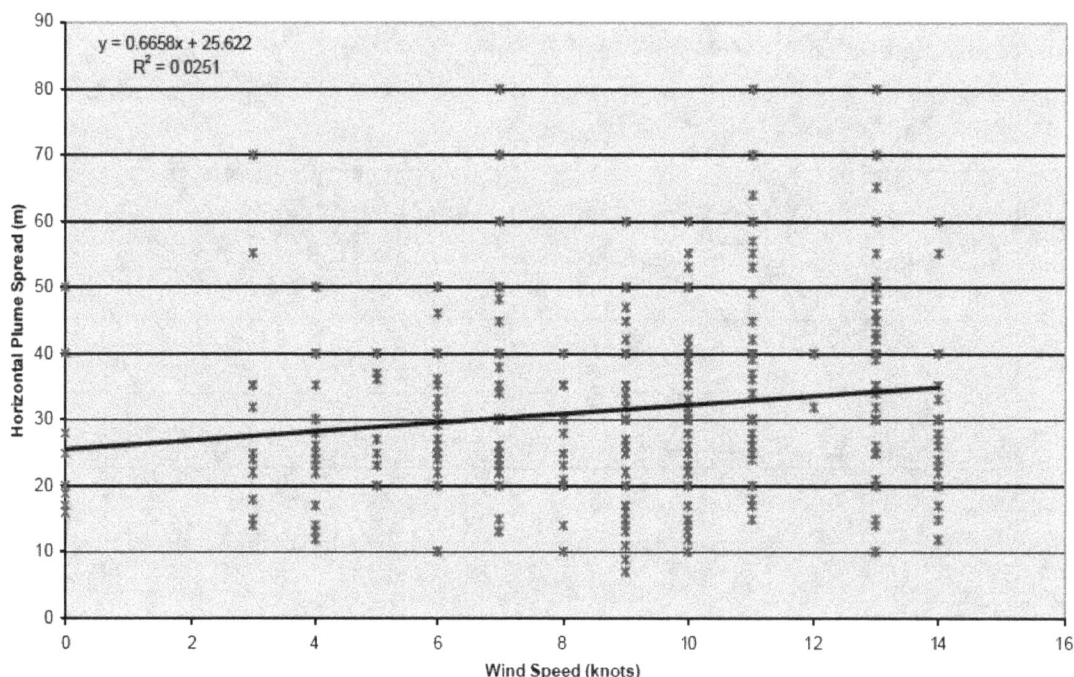

Figure 11. Large Commercial Aircraft Initial Plume Parameters Vs. Wind Speed
(continued)

c. Wind Speed Vs. Vertical Plume Spread

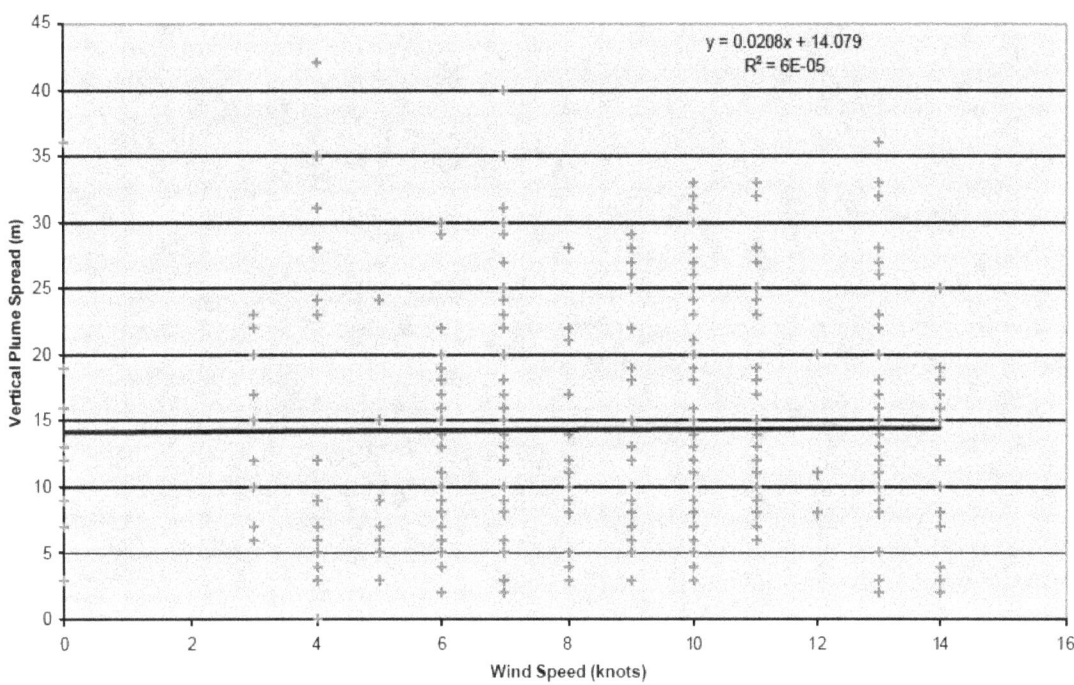

Figure 12. Commuter Aircraft Initial Plume Parameters Vs. Wind Speed

a. Wind Speed Vs. Plume Rise

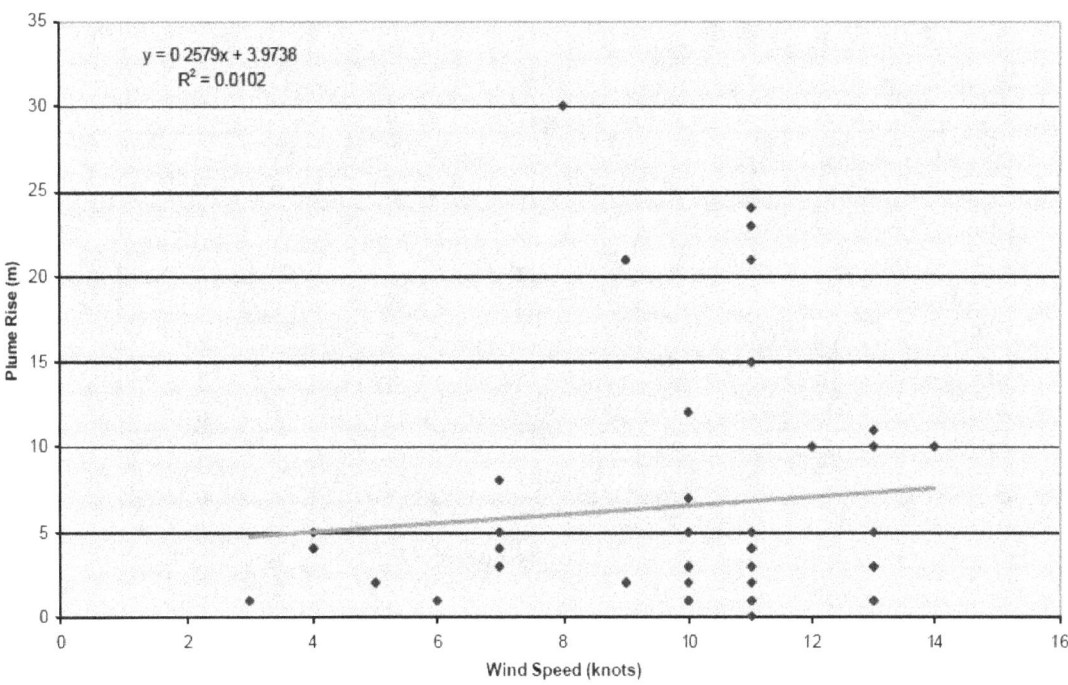

Figure 12. Commuter Aircraft Initial Plume Parameters Vs. Wind Speed (continued)

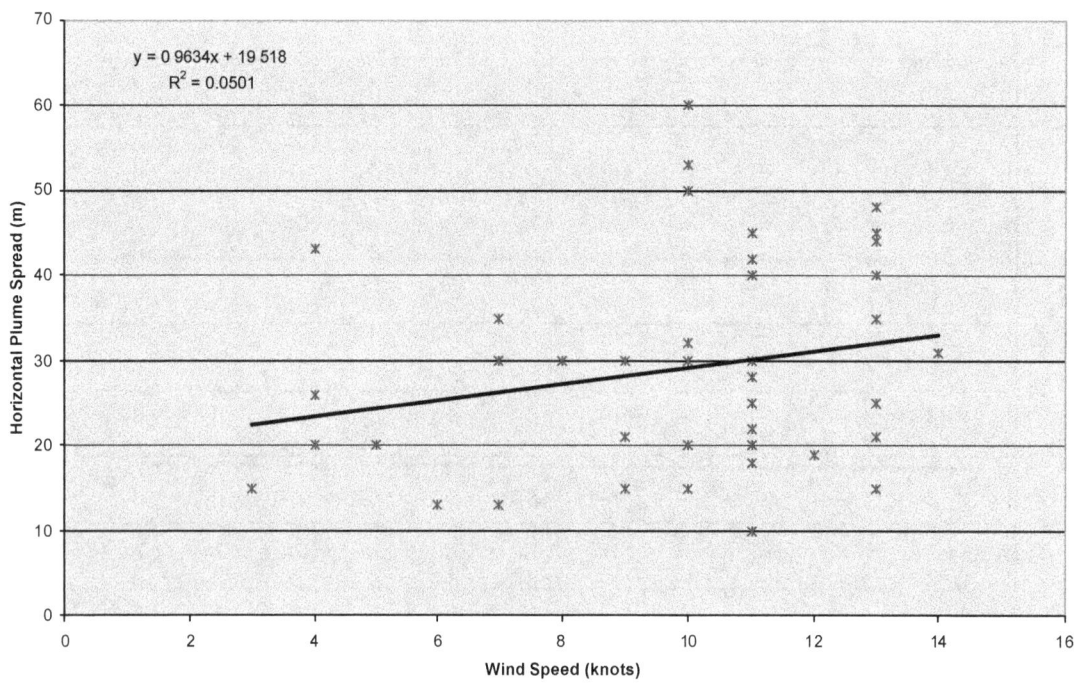

b. Wind Speed Vs. Horiz. Plume Spread

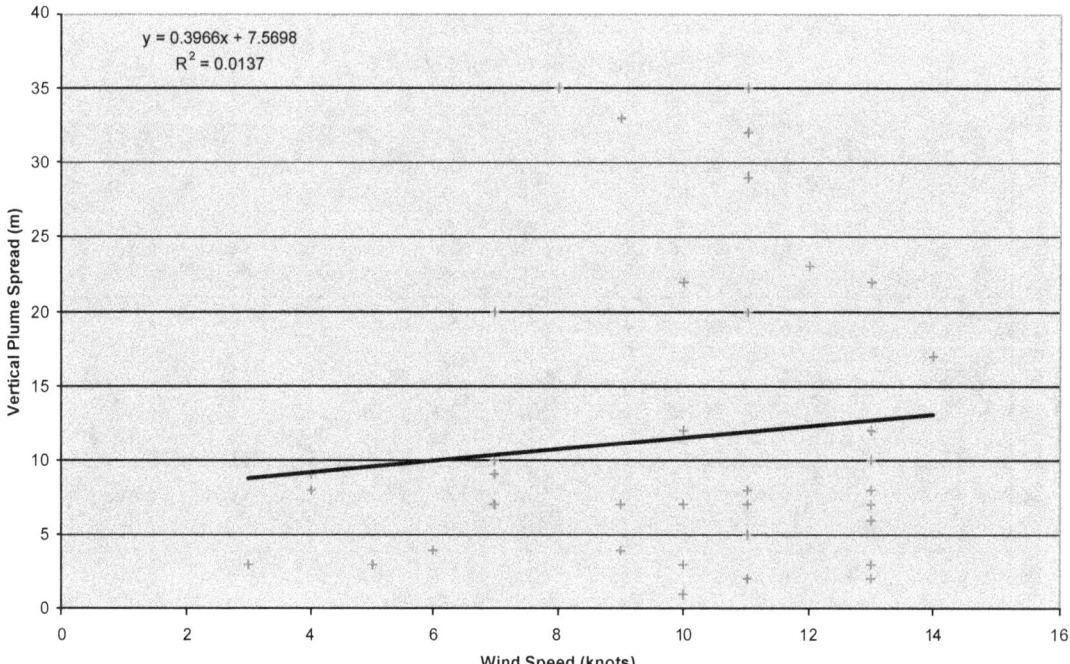

c. Wind Speed Vs. Vertical Plume Spread

Figure 13. Large Commercial Aircraft Initial Plume Parameters Vs. Wind Direction

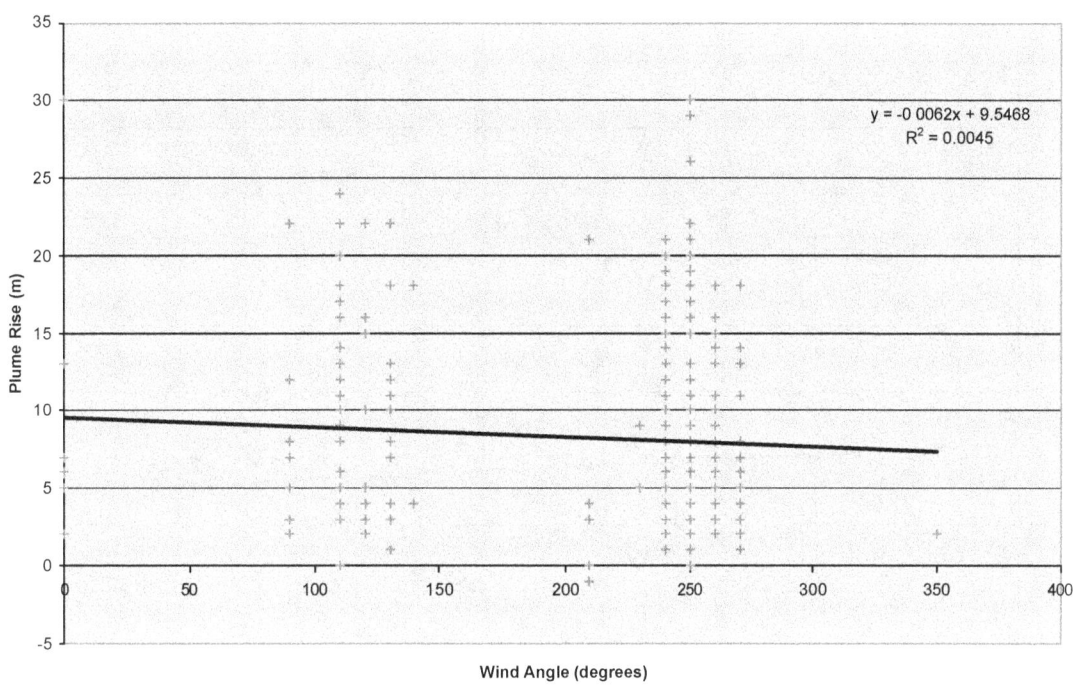

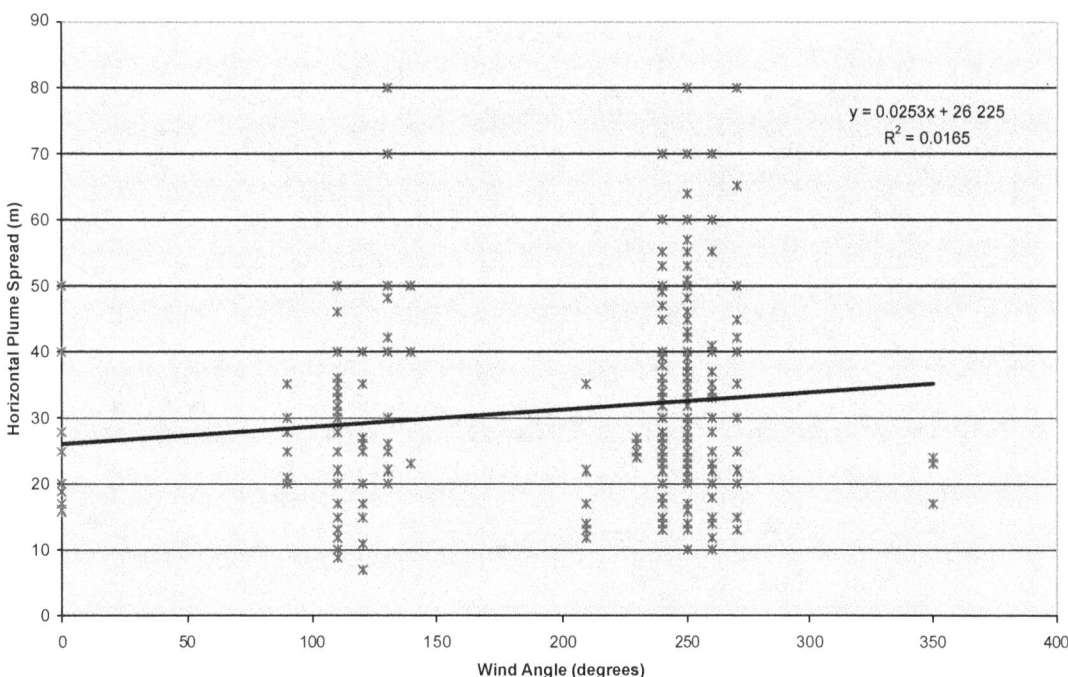

Figure 13. Large Commercial Aircraft Initial Plume Parameters Vs. Wind Direction (continued)

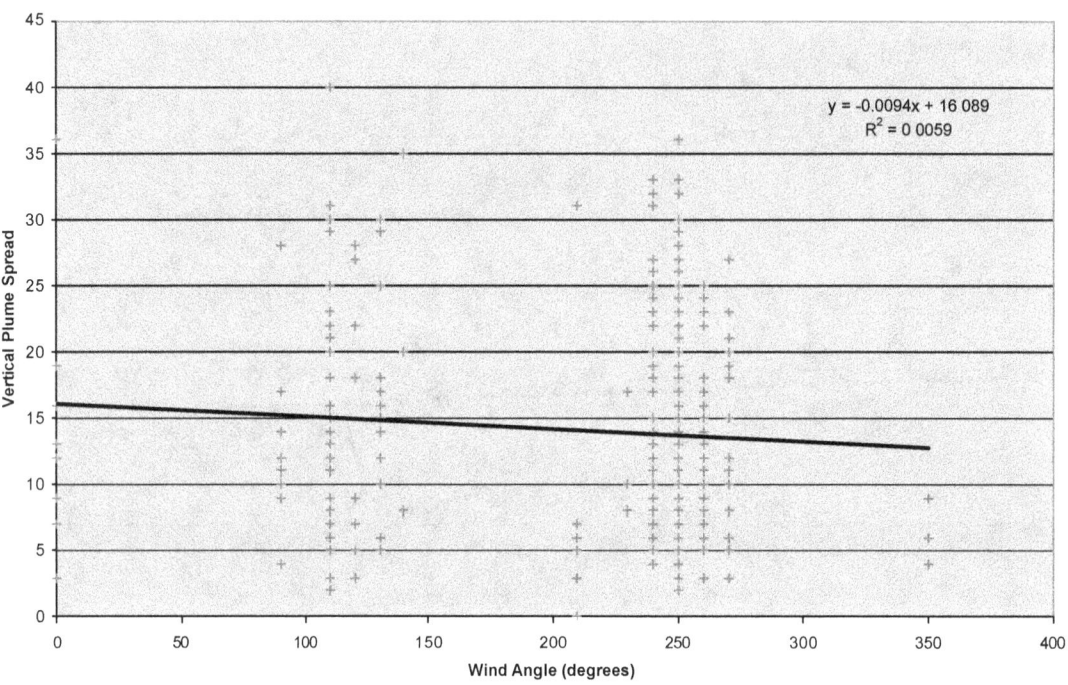

Figure 14. Commuter Aircraft Initial Plume Parameters Vs. Wind Direction

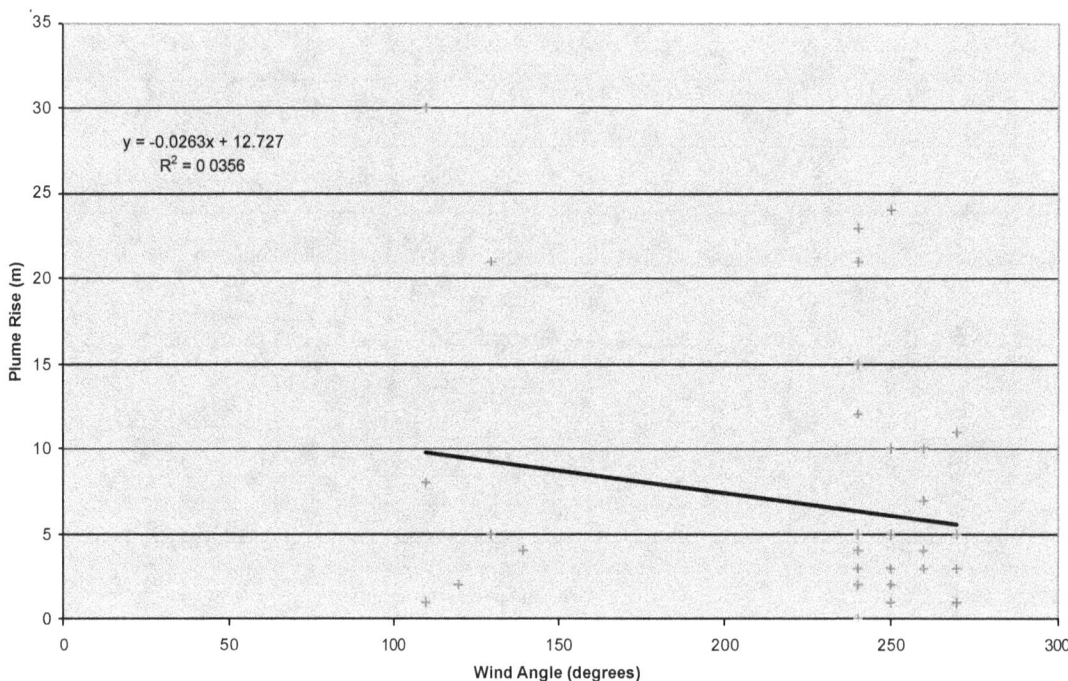

Figure 14. Commuter Aircraft Initial Plume Parameters Vs. Wind Direction (continued)

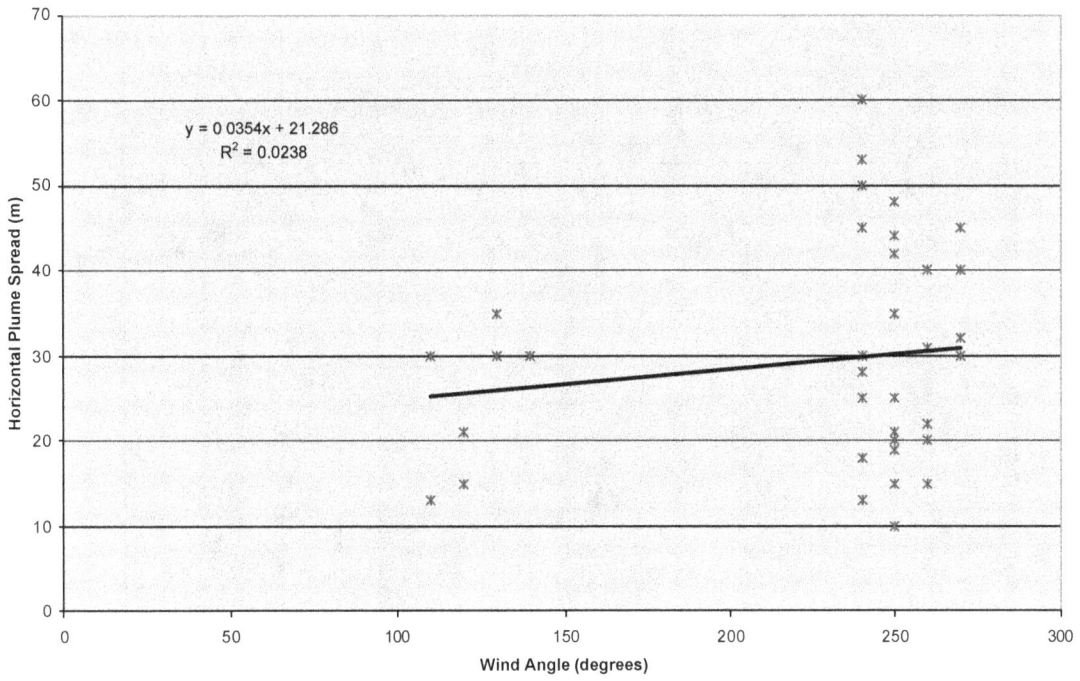

b. Wind Direction Vs. Horiz. Plume Spread

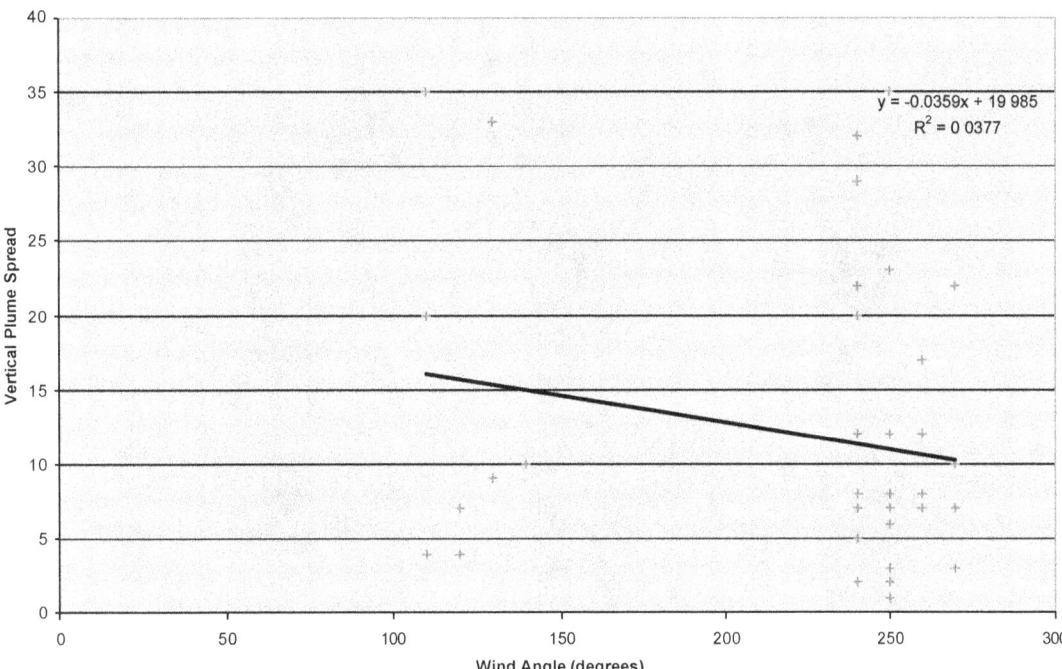

c. Wind Dir. Vs. Vert. Plume Rise

Figure 15. Comparison of Initial Plume Parameters Vs. Stability Classes

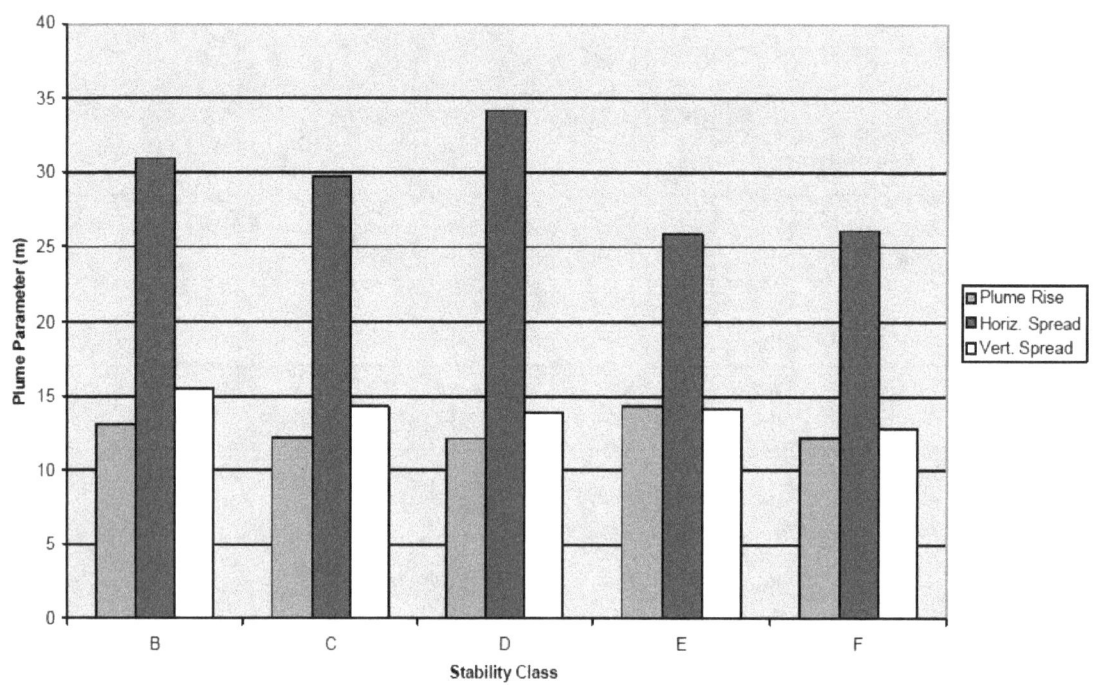

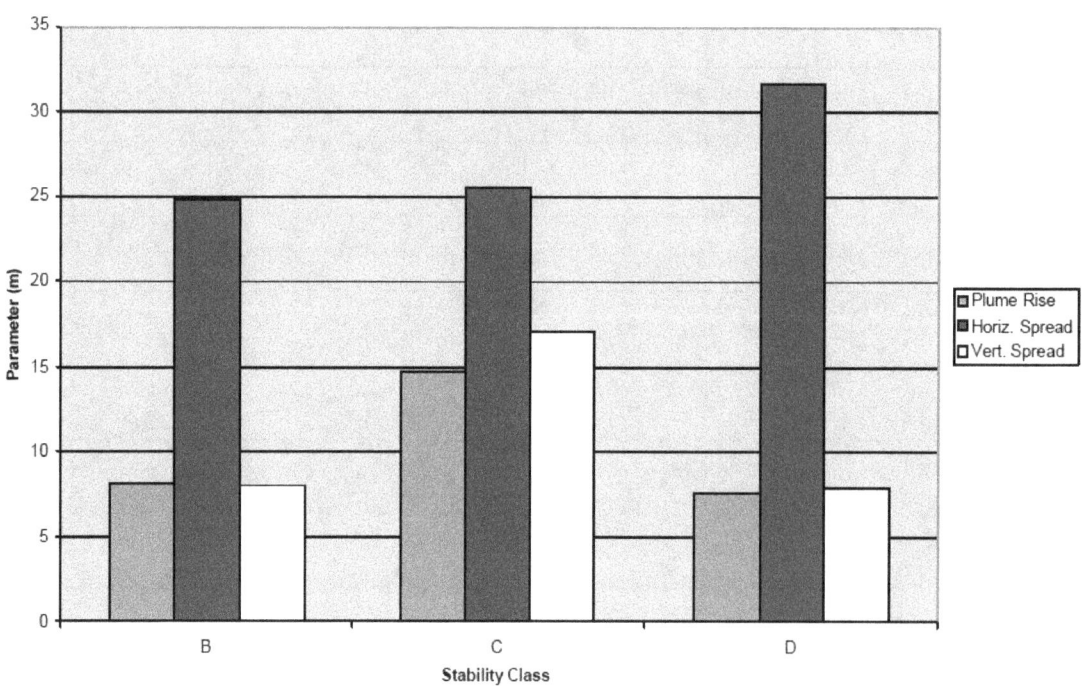

DATA RESULTS

This study provides new data for more accurate modeling of plume rise and spread. Findings in this report represent overall values for plume rise and initial plume standard deviations. Based on these findings the final overall values, based on the aggregate analysis are:

Large Commercial Aircraft (primarily turbofan engines)

    Sigma Y =   10.8 meters
    Sigma Z =   4.1 meters
    Plume Rise = 11.9 meters

Commuter Aircraft (primarily turboprops)

    Sigma Y =   10.3 meters
    Sigma Z =   4.1 meters
    Plume Rise = 12.1 meters

Due to the similarities in these preliminary values, it is recommended that a single set of values be utilized until the follow-on analyses at additional airports are complete. The single set of derived values is as follows:

    Sigma Y =   10.5 meters
    Sigma Z =   4.1 meters
    Plume Rise = 12 meters

CONCLUSIONS

General

This study provides new insights into aircraft plume behavior that greatly surpass historic understanding, and data for more accurate modeling of plume rise and spread. It can be concluded that significant plume rise occurs for the jet/turbine exhaust plume. It can also be concluded that initial plume spread is significant and greater than previously thought. Since there is only one data set, study results for plume rise and initial plume standard deviations are calculated using a conservative basis (i.e., the plume was only measured to well-defined boundaries, the plume rise was based on the sweep corresponding to the second greatest height or elevation measured, the sweep used for the plume rise was also used to determine the standard deviation of the plume). Findings in this report represent aggregate values for plume rise and initial plume standard deviations. Effects of aircraft type, temperature, wind speed, wind direction, and turbulence (stability class) were not found to be statistically significant in the data analysis. While differences do appear to occur by aircraft types, it cannot be proven they are significantly different at this time and more measurements are needed to follow up on this trend analysis.

Plume Rise Dependence on Thermal Buoyancy

The results of the measurements would tend to support that jet exhaust plume rise occurs due to the plume's initial thermal buoyancy. Plume rise has been studied intensely for stationary sources such as stacks and semi-empirical equations have been developed and successfully applied in predicting plume rise. The important variables used in these approaches often include the wind speed, downwind distance, and heat emission rate.[7] This allows many of the derived models to take the form:

$$\Delta h(x) = \text{constant } (Q_h)^a (x)^b (u)^c \qquad [2]$$

where:
- $\Delta h(x)$ = plume rise as a function downwind
- Q_h = heat emission rate
- x = distance downwind
- u = wind speed at source height
- a,b,c = constants

The derived stationary approach cannot be used in this case because most of the equations include the vertical velocity of the release, common for stationary sources. Due to the horizontal release of the exhaust gas from the aircraft engine, vertical momentum/vertical kinetic energy is small when compared to thermal buoyancy. If the jet exhaust is considered to have the properties of air, the thermal buoyancy is a function of the difference in the absolute temperature of the jet exhaust and the ambient air. The internal temperature of the primary combustion zone of a modern jet/turbine engine can reach 2000 degrees Kelvin. However, the temperatures at the turbine are closer to about 1300 degrees Kelvin and as such, the exhaust stream from the core engine will be in the range of 1000 degrees Kelvin or greater. This temperature is reduced considerably by mixing with the ambient air and the bypass air of modern turbofan engines. Regardless, the exhaust temperature of the engine is still very much above the ambient temperature. This leads to a large heat emission rate and as such, a large amount of thermal buoyancy.

The distance downwind where plume rise stops is also an important factor. In this study, this distance was approximated since plume rise was documented by having multiple LIDAR sweeps across a fixed plane as the aircraft continued its taxi or takeoff roll and the second maximum value was selected. For example consider the aircraft event presented in Figure 16. This is a histogram of the measured plume rise for each sweep of a DC9 aircraft on takeoff. The plume continues to rise and reaches a peak. Then, as the aircraft continues down the runway, the plume reaches a peak and begins to dissipate.

Notice that the second highest value was selected in this conservative approach to estimating plume rise. This conservative approach helps to insure under-prediction of

---

[7] Good reviews of this material can be found in:
Stern, A.C. Ed., Air Pollution, 3rd Edition, Volume I, Academic Press, New York, 1976.
Zannetti, P., Air Pollution Modeling, Bookcraft Ltd, Avon, U.K., 1990.

local ground level concentrations will not occur. From the analysis of the change in an "aged" plume behind the aircraft, the distance downwind is accounted for.

Wind speed effects, including the direction to add a vector consideration of the wind in the analysis, were also analyzed in this work as reported in the last section of this paper. However, as shown, no correlation was found for this data set, inferring that wind speed was not an important variable.

Figure 16. Plume Rise Trace by Time Dependent Sweep

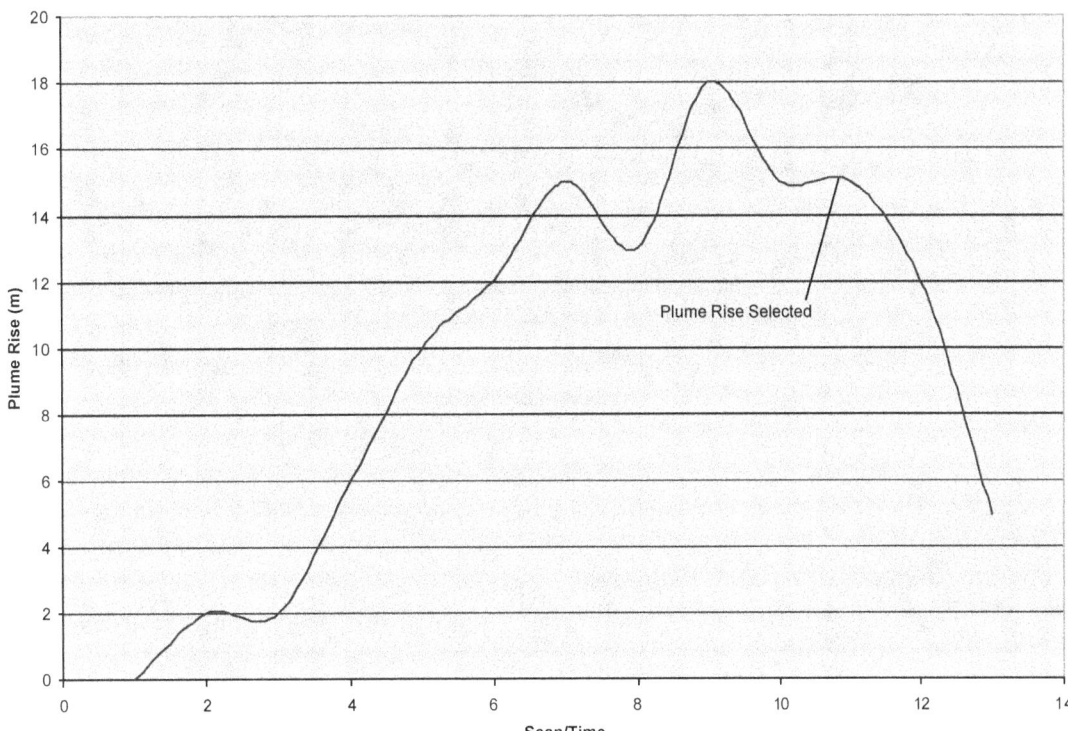

This would mean that Equation 2, for our particular case, could be reduced to:

$$\Delta h(x) = \text{constant} \, (Q_h)^a \qquad [3]$$

Of course this implies that thermal buoyancy is the overriding variable for the plume rise of a jet/turbine engine exhaust. Intuitively, this makes sense. However to check this conclusion more closely, the derived initial plume parameters were analyzed once again. This time, the initial plume spread was compared to the plume rise. Figure 17 shows this analysis for large commercial aircraft while Figure 18 shows the analysis for the commuter aircraft.

Figure 17. Large Commercial Aircraft Plume Rise Versus Plume Spread

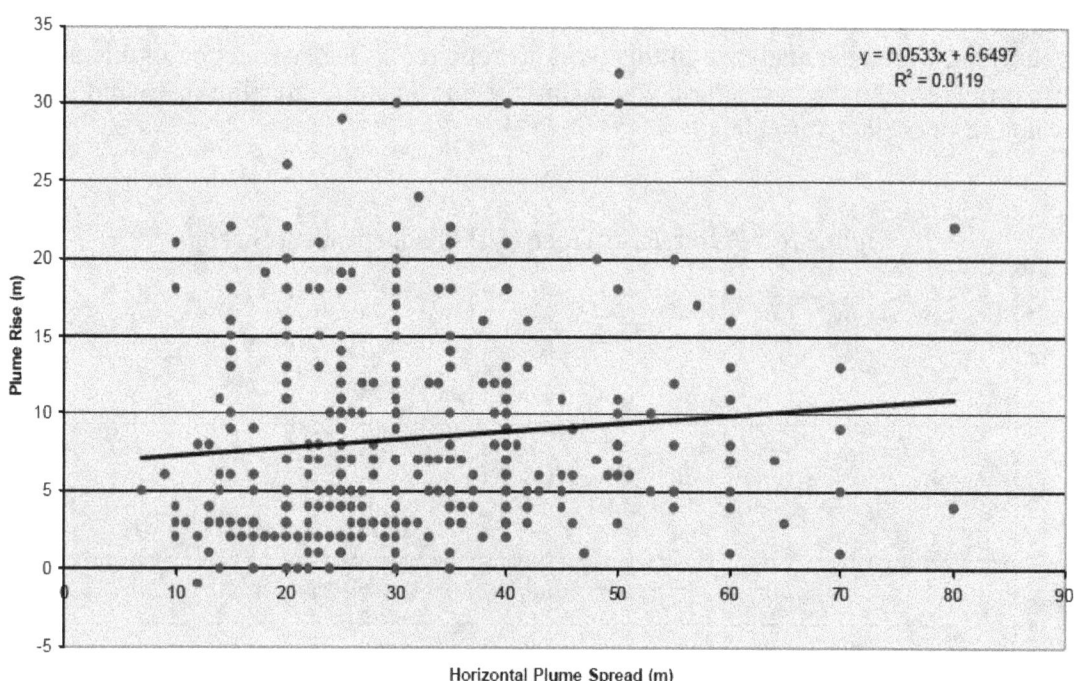

a. Plume Rise vs. Horiz. Plume Spread

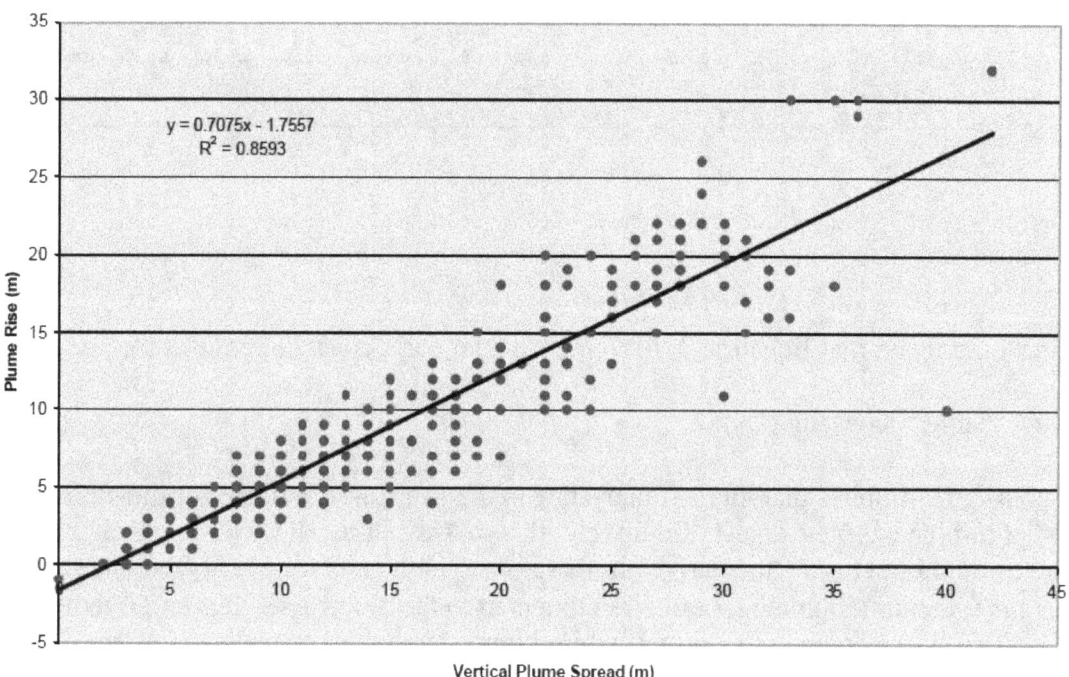

b. Plume Rise Vs. Vertical Plume Spread

Figure 18. Commuter Aircraft Plume Rise Versus Plume Spread

a. Plume Rise vs. Horiz. Plume Spread

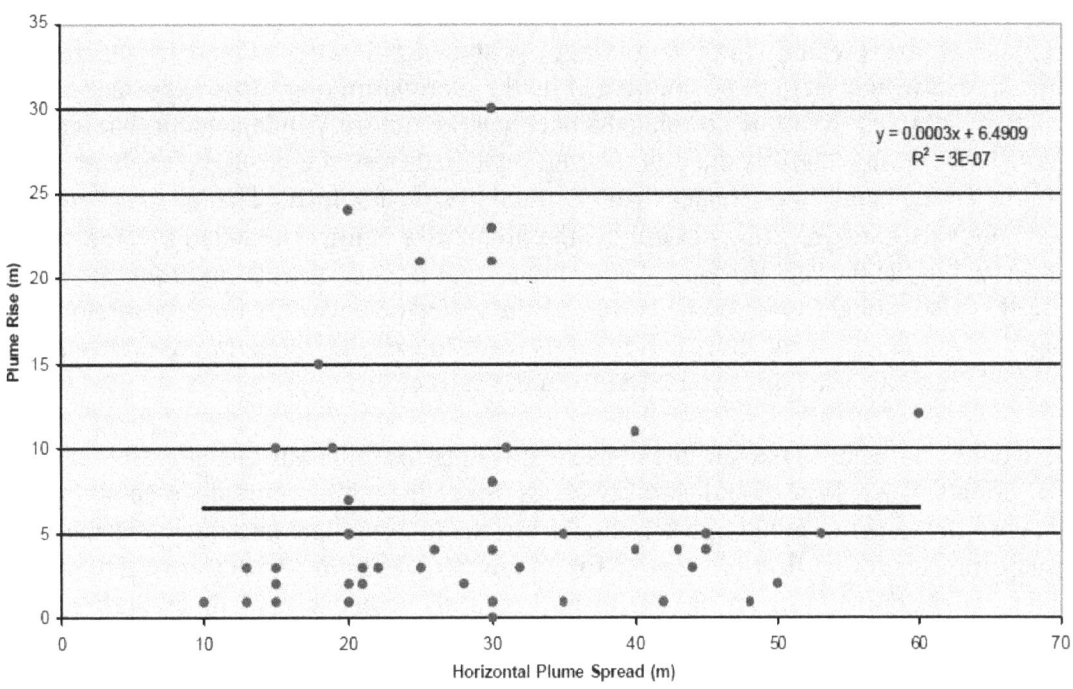

b. Plume Rise Vs. Vertical Plume Spread

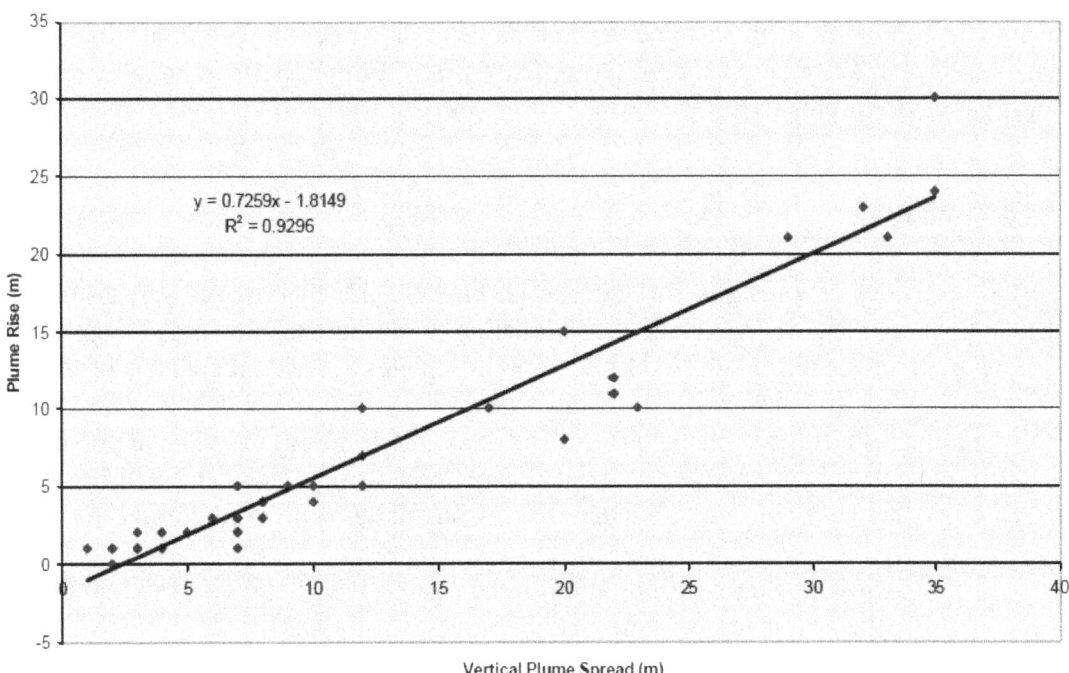

As can be seen from Figures 17 and 18 the horizontal plume spread does not correlate with plume rise while the vertical plume spread shows excellent agreement with the plume rise. Plume spread is a function of the atmospheric mixing which is a combination of mechanical and thermal mixing. Vertical plume spread is more greatly affected than is horizontal plume spread by thermal mixing. As shown in Figures 17 and 18, the vertical plume spread would seem to be dominated by the thermal mixing. It is expected that these two variables would be correlated since they are not truly independent, but it should be noted that during analysis the plume often separated from the ground plane and as such, the plume height was greater than the plume vertical spread. That so close of an agreement exists suggests that not only is the plume rise being dominated by thermal buoyancy, but the thermal buoyancy is having a large impact on vertical plume spread as well. It is also thought, based on a review of the collected data, that the core engine exhaust does not mix completely and that this leads to extreme temperature differences and rapid plume rise causing separation from the surface.

This would also seem to explain the consistent results from the various aircraft and the lesser effects of wind speed and stability class. Since the temperature differential (exhaust – ambient) is quite large for all aircraft, the thermal buoyancy is by far the primary impetus for vertical plume rise for all aircraft. The consistency between aircraft could exist because all aircraft have temperatures at the exhaust plane much higher than ambient and as such, all have a strong upward force due to the thermal buoyancy. This would also explain the wind speed being slightly more correlated with horizontal plume spread than with vertical plume spread. It should be noted that is a relative statement, based on this data set, since the correlations were so low for all wind speed correlations.

The finding from this data set that plume rise may only be dependent on the thermal buoyancy is important. If this can be verified as more data becomes available, the estimation of plume rise could follow the simple form of Equation 3 and be related to exhaust gas temperatures only.

FUTURE WORK

This study provides new insights into aircraft plume behavior that greatly surpass historic understanding, and data for more accurate modeling of plume rise and spread from commercial aircraft at airports. This final report completes individual analysis of the LAX data set, initially reported in the related *Preliminary* Report published in September 2002. Additional studies are planned (based on available funding) to evaluate potential changes in the derived parameters due to site characteristics (site bias) and will be reported on as the work continues. These follow-on studies will be important to:

- Allow further exploration of plume rise variables and confirm that the exhaust gas temperature may be directly correlated;
- Provide a more extensive data set to allow the analysis of the effects of aircraft types and atmospheric stability to be more accurately defined;

- Collect more detailed weather data, collected event by event rather than the NOAA average values to allow more detailed evaluation of weather effects.
- Explore the determination of the mass of aerosols across a plume and the relationship to particulate matter emission indices by conducting concurrent sampling of aerosols near the aircraft to "calibrate" the LIDAR results.

Future work should also be done at other altitudes, during colder temperatures, at a greater range of wind speeds, and more data should be taken during stable atmospheric conditions. Each data set will add to the body of knowledge and improve the understanding of plumes emitted by aircraft.

www.ingramcontent.com/pod-product-compliance
Lightning Source LLC
Chambersburg PA
CBHW081808170526
45167CB00008B/3382